Henri Bergson, David Bohm, and the Holographic Theory of Mind

A Bergson-Bohm Comparison

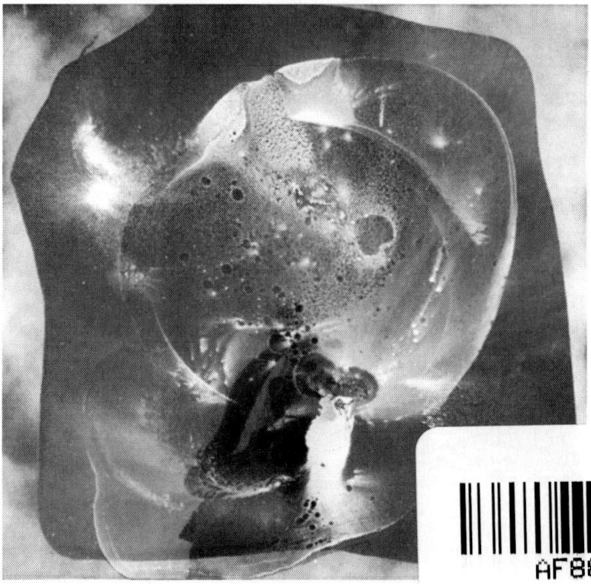

A Dialogue between

Michael Pryzdia and Stephen Earle Robbins

By Michael R. Pryzdia

The Viola Institute

Dedicated to Henri Bergson and David Bohm for the numerous key ideas brought forth in their work.

Published in the United States of America

Printed in the United States of America

Henri Bergson, David Bohm, and the Holographic Theory of Mind: A Bergson-Bohm Comparison

ISBN-13: 978-1-950761-30-2

"There must be no opposition between time and the Timeless."
Jiddu Krishnamurti

"Meditation is the moving without any past."
Jiddu Krishnamurti

CONTENTS

A Dialogue Between
Michael Pryzdia and
Stephen Robbins

Introduction

Is Henry Bergson's Holographic Theory in alignment with David Bohm's Holographic Theory?

It was during the Christmas Holiday season of 2022 that Stephen Earle Robbins and I first sat down and recorded a series of Talks on Zoom. A couple of weeks prior to the recordings, I had asked Steve if he would be interested in doing a few informal presentations on a handful of topics that I was interested in getting his opinion on. I was reacting to his notion that the French philosopher Henri Bergson had put forth a "Holographic Theory" in his groundbreaking work *Matter and Memory* (1896). Being a fan of the work of the quantum physicist David Bohm, who had his own holographic theory as put forth in *Wholeness and the Implicate Order* (1980), I soon became somewhat obsessed with the project of integrating the work of Bohm and Bergson.

As I share in Talk 1, I was initially shocked by much of what Steve was saying in his own work, but his YouTube videos, books, and articles eventually infected me; and I had to admit to myself that there was much value in what he was sharing. I was already researching and teaching within the relatively new academic field of "consciousness studies" and was somewhat familiar with J.J. Gibson and Karl Pribram and what they had to say about the nature of perception. And I was quite interested in how Steve was ricocheting off of what Pribram was saying by integrating Gibson and Bergson.

Eventually, I reached out to Steve and attempted to share a few of the ways that Bohm was influenced by the teachings of Jiddu Krishnamurti. And, to make a long story short, Steve and I opened up a dialogue centered on my obsession with integrating Bohm and Bergson; for it was obvious to me that while both Bohm and Bergson had holographic theories, the nature of those theories, while very similar, were also very different. For example, each thinker had quite different ideas regarding the nature of perception, memory and novelty/creation. The similarities and differences were what motivated me to ask Steve to do a few presentations regarding "Bergson and Bohm." In the end, over the course of a few weeks, four Zoom presentations were created which enabled Steve and I to dialogue with each other.

These Zoom presentations were then edited and transcribed. The four talks and transcripts are shared here:

Bergson and the Holographic Theory of Mind –
A Bergson-Bohm Comparison
(Talk 1: "On Bergson and J.J. Gibson")

Bergson and the Holographic Theory of Mind –
A Bergson-Bohm Comparison
(Talk 2: "On Bergson, Pribram, and Bohm")

Bergson and the Holographic Theory of Mind –
A Bergson-Bohm Comparison
(Talk 3: "On Bergson and Bohm, Part 1")

Bergson and the Holographic Theory of Mind –
A Bergson-Bohm Comparison
(Talk 4: "On Bergson and Bohm, Part 2")

This work will most likely most appeal to individuals interested in the work of David Bohm, Henri Bergson, and Consciousness Studies in general; but because many other issues are alluded to, issues connected to quantum physics, relativity, perception, memory, and creativity, many others will be able to garner some useful information connected to their research and/or personal interests.

My goal in asking Steve Robbins to do all of these presentations with me was not so that I could share how much I am in agreement with what he shares. On the contrary, my intention was to have an honest-to-goodness dialogue with him, so that he and I together could do some exploring and hopefully make some discoveries – which I feel we did do. I realize that in addition to what Steve is saying, some people may be put off by his tone and his presentation style – as I was when I first discovered his work. But I think these presentations are important because Steve does have some very important things to say – and, in many respects, we see Steve at his finest here. The fact that the academic community has virtually ignored his work up to this point in time reminds me of how Bohm was himself ignored in his day (i.e., the now infamous Oppenheimer quip: "If we cannot disprove Bohm, let us continue to ignore him.").

In future writings I am hoping to further examine what is shared here so that those interested in exploring ways in which Bergson and Bohm may (or may not) be able to be integrated can be rewarded. That is my intention. Do enjoy!

Michael Pryzdia – Phoenix, Arizona

Talk 1: On Bergson and J.J. Gibson

Pryzdia:

Okay. . . Hey Steve, are you there?

Steve:

Yep, I'm here.

Pryzdia:

Well thanks for agreeing to do this with me. I've been looking forward to this. I appreciate you taking the time out to take us through . . . What are we calling it? "Bergson's Holographic Theory -- with a focus on Bohm." As you know, I am a big fan of Bohm's work; so, I am looking forward to this. And I think I should share the story about how we actually met. As you may recall, I think it was back in 2017, this is when you had just finished making one of your YouTube videos – which was part of your series on "Bergson's Holographic Theory." And I saw the one on Bohm -- and was appalled. So do you remember how we had this YouTube commentary exchange where I'm sitting there watching this presentation you did on Bohm where you are comparing and contrasting Bohm and Bergson – and, by the way, at that time I hadn't read any Bergson at all – and I wrote you. And we went back and forth and it took me about . . . I think it took me about a year to assimilate what you are actually saying about Bohm and Bergson in your comparison. So, I went back to your presentation and I reviewed it maybe twice, and I'm absorbing this thing. And I'm saying, "Jesus, I can't deny what Steve is saying is actually correct. There are some really good points here." So, it forced me to look at Bohm from a new perspective (ala "Bergson's Holographic Theory"). And by the way, I never thanked you for turning me on to Bergson because I have become a bit of a Bergson freak. So, there's a lot of good stuff here, and I have a hunch that other people who find your videos on YouTube and find your Bergson's Holographic Theory presentations . . . I think a lot of people might come to it by way of Bohm. And some of these Bohm people might have the same reaction initially that I had. So, I think this is a good idea.

Steve:

I got the distinct impression that what I was saying was going right past you at the time.

Pryzdia:

Exactly, exactly! It took a while to absorb what you were actually saying. It's almost like . . . well it's like what Bohm used to say: people don't really listen to each other with conscious awareness. It took me a good year to even have the courage to go back and re-watch the presentation. Obviously, there was something you said that hooked into my brain and it took me a while to say, "Hey, there's something there. There IS something there." So, I went back and reexamined your presentation. By the way, I do want to say though that maybe today or the next time we talk, I could bring up some of the initial points I brought up back then in 2017, or maybe it was 2018 -- because we never have had a conversation about Bohm's relationship to Jiddu Krishnamurti and how Krishnamurti's ideas very much influenced Bohm's physics. I remember in my interchange with you, we were talking about the term "direct perception" -- because it's a term that Bergson uses in one way, but it is also a term that Bohm used with Krishnamurti -- but they used it in an entirely different way. One of the things you said to me was, "Bohm doesn't have a theory of perception." And I remember writing to you: "What do you mean?! He talked about that all the time with Jiddu Krishnamurti." So, it's a question of language here, but I think at some point it would be prudent and interesting to go back and maybe bring in some of what Bohm learned from Krishnamurti around perception, not in any technical sense. That is, he did not have a technical "theory of perception"; I agree with you there; but there's a lot of gold to be mined there with regard to what they were saying about perception.

Steve:

Okay, there we go. Well, there it is:

Bergson and the Holographic Theory of Mind

The question I think you want to ask is: "How did I ever discover Bergson in the first place?" And that's where it gets sort of interesting. Truthfully, I had been in Vietnam and when I'd come back in North Carolina, I ended up having, shall we say, a mystical experience after . . . well, the war experience makes you very profoundly involuted in looking at the meaning of life, et cetera, et cetera. But this particular experience tuned me up to the question of subject and object, believe it or not. And curiously enough, almost very quickly after that experience, I was released from the military and made my way to the University of Minnesota where I was working on a PhD in educational psychology. That was my intent. I had a background in psychology and political science before Uncle Sam grabbed me. But I also loved Piaget at the time from my previous background, before the war; I had discovered Piaget in cognitive science.

So, I was taking a course on Piaget and group theory, and the professor, a Dr. Bart, had an assignment of course and wanted us to write a paper. And I said, well, I'm going to write a paper on Piaget and his epistemology. Now you might ask, where did the epistemology fly out of? And that was from courses in philosophy that I had taken that had interested me, but I had no clue what epistemology really was all that much about. But anyway, he said, well, why don't you write a paper on the relationship of Piaget and Bergson? And I said, "Oh, okay." I'm thinking, "Who's Bergson?" Same as everyone else: "Who is Bergson?" So, I found a book **on** Bergson – because one never reads Bergson directly right off the bat. You find someone that's kibitzing about Bergson. And this was a book by Herbert Wildon Carr (*Henri Bergson: The Philosophy of Change*, 1911). But what struck me right away was . . . Let me see if I have the quote here.

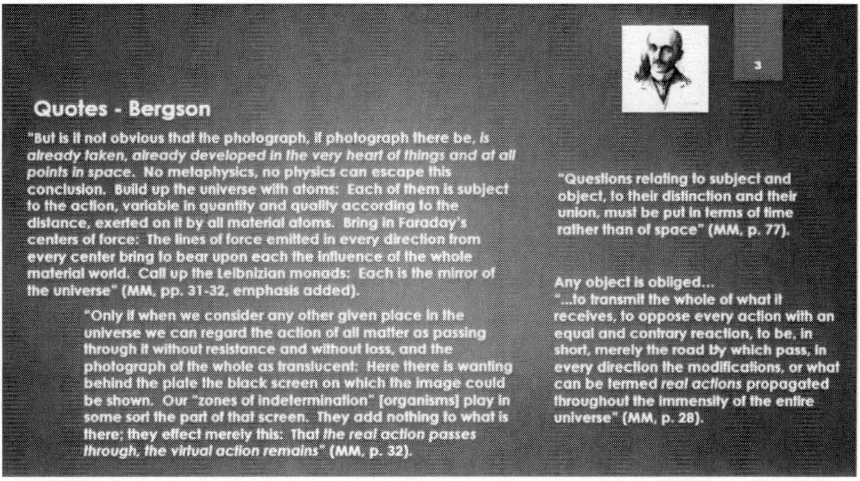

I have my quotes from Bergson up there -- that little question, the little quote over on the right side that states that: "Questions relating to subject and object, to their distinction and their union, must be put in terms of time rather than of space (*Matter and Memory*, p. 77)." And I'm going . . . "What the heck does that mean?" It was like a Zen Koan – I'm thinking, "What does he mean by that?" This was a quote from Bergson's *Matter and Memory* book that Carr was reiterating. And that is what made me realize that there was something rather profound going on here. And I started reading *Matter and Memory*. Now, the experience one has when one reads *Matter and Memory* -- which was written in 1896 -- is that it is unbelievably difficult to comprehend. There are four chapters; and Bergson was afraid of the following precise phenomenon – that Chapters One and Four are very philosophical; Chapters Two and Three course through the then current memory research of the day -- 1896. And Bergson was intent to ask the question: Is memory stored within the brain? And he considered that question to be the **key** question for a theory of consciousness. And it is. But people don't understand that yet -- to this day. If you said: Is memory stored in the brain the key with regard to the problem of consciousness? -- And you said that to Chalmers, he would not have a clue to what you're asking. But that was what got me going, that question – "the subject and object question." And obviously as you read the book, you then realize that there's a whole theory of a whole new theory of the brain, of memory, of perception. And, of course, the problem with the perception theory was that it incomprehensible. And it wasn't only incomprehensible to me; it turns out that it was considered extremely obscure by all the people that read the book of his own day.

So that's how I got going. And then at the end of 1970 I started graduate school; and beginning in 1972, the whole computer model of mind swept into academia. And my advisor then, Paul Johnson, and the educational psychology side of things, was extremely interested in the book by Newell and Simon called *Human Problem Solving* (1972). And we did a seminar on that. And of course, Newell and Simon were trying to simulate human problem solving. That was the initial state of affairs when this whole phenomenon swept in -- this whole movement of the mind as a computer. The state of affairs was that we are going to attempt to simulate or show how the brain actually works: it is a symbolic processor. And that, of course, has been nearly abandoned over the current years. And AI makes no pretense. The AI folks would like to be able to do what people are doing, but as far as simulating how people actually do it, that's almost considered a lost cause. But simultaneously, I was taking seminars in the psychology side of the University of Minnesota with Bob Shaw; and Shaw was a student of Gibson. He had been with Gibson at Cornell – J.J. Gibson, the great theorist of perception. Shaw had been at Cornell for a couple of years with Gibson. He was brilliant. He was also heavily versed in computational modeling. And in theoretical, abstract automata theory, et cetera. In fact, in the seminar, he went through one of these books, I think it was by Feigenbaum -- *Computers and Thought* (1963). It was a whole series of chapters on various computer programs that were trying to simulate mind. And I mean, this was already in 1972, and he just annihilated all of them as not even close to dealing with the problem. Which if you look back at it and realize this is already 1972 and here we are -- 50 years later, and we still haven't caught up with Shaw on that. Of course, Shaw was a Gibsonian. And so now you're also bringing in Gibson's theory of direct perception.

I think the interesting thing here is that we get to Pribram very quickly in this story; because Pribram came to visit University of Minnesota and visit Shaw. This was again, roughly maybe 1974. And Shaw obviously had discussions with Pribram. And Shaw gave a seminar on one of the days that Pribram was there. And Shaw started talking about holography. And he described how holograms work.

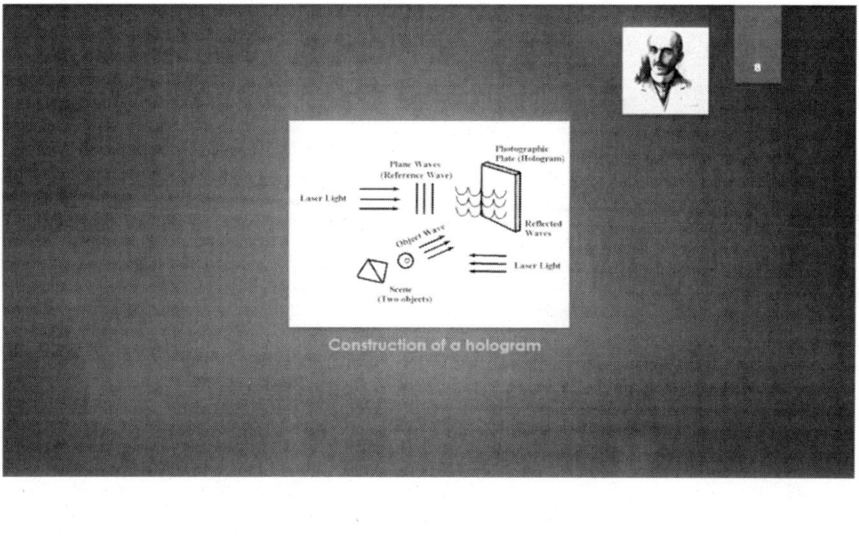

He would have said something like this (see above): when you're constructing a hologram, you've got an object, like in our little picture there -- a pyramid and a ball. And you've got a laser source, a laser light, and you're sending the laser light toward the hologram plate there. And sometimes there's a half mirror that sends another part of the wave down to the object wave, and it bounces off the objects and also hits the plate. So, what you have is the intersection of two waves at the hologram plate. And if you look at a point on one of those objects, just a point -- you see every point is giving rise to a spherical wave that's traveling toward that plate and covering the entire hologram. So that means that the information for any point of the object, say of that ball, is covering the entire plate. But there's a converse. And that is that at every point in the plate, therefore, is the information for the entire object, for the entire ball and the pyramid. So, each point, shall we say, is the reflection of the whole, has information for the whole.

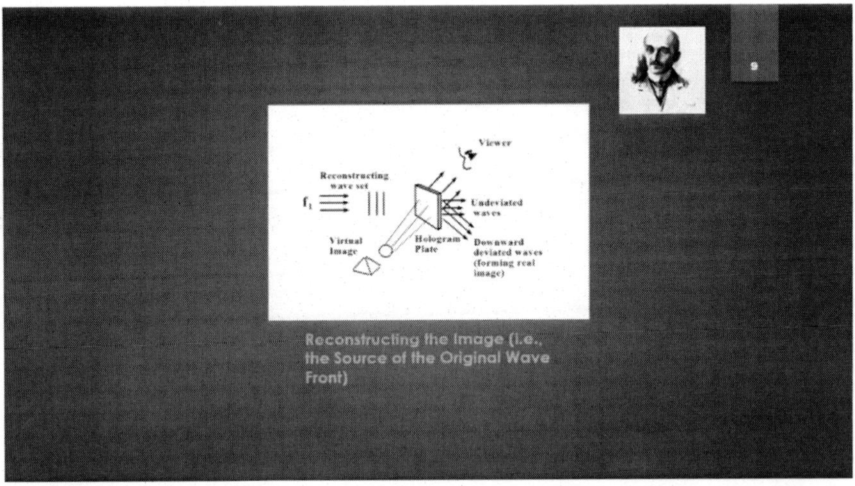

Now (see above) when you send the reconstructive wave back through the plate (and it's got to be the same frequency as the original reference wave, the wave that you used initially from the laser), you send that through and the viewer sees the virtual image in 3-D of the original object, namely the object wave that was coming from the pyramid and the ball. So that's nice.

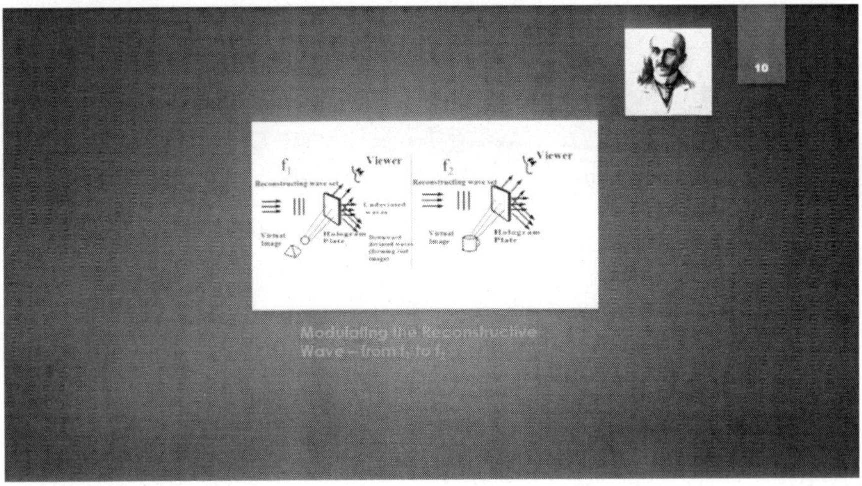

And we go to another picture (see above), you can see that we can modulate the reconstructive wave. For example, I could have used – let us abstractly call it *frequency 1* -- as a reference wave from the laser going through the plate, and recorded the wavefront, the interference pattern of the pyramid and the ball. Or I could have used *frequency 2*, a different frequency, and recorded a wavefront interference pattern from the cup. But now all I need to do -- as I **modulate** that reconstructive wave -- if I modulate it to *frequency 1*, I see the image of the pyramid and the ball. If I modulate to *frequency 2*, I see the cup. So that's the essence of holography. These principles of modulating the reconstructive wave. And of course, the idea is that each point on the hologram is a reflection of the whole, that it carries information for the whole.

So, Shaw was talking about Pribram, and of course Pribram thinks that the hologram is the brain. Pribram had already written his book called *Languages of the Brain* in 1971, and that was his model. The brain was a hologram, and he was trying to figure out how interference patterns, maybe some neural flows or neural patterns, were interfering with each other in the brain; and the brain was the hologram. And Shaw said, and this was kind of an amazing "drop the bomb type statement," "What if I were to tell you that we are in the hologram; the universe is the hologram **and the brain is the modulated reconstructive wave passing through the hologram**. In other words, the hologram isn't so much in the head as the head is in the hologram." That was his phraseology then. And at that point, I guess I must have had an "aha" experience because I remember telling Shaw after the seminar, "that's Bergson's model -- that's what he's saying." And how did I know that? Well, by that time, I'd probably read Bergson's *Matter and Memory* a dozen times -- inside and out, from the front to the back, from the back to the front -- trying to grasp what he was really saying.

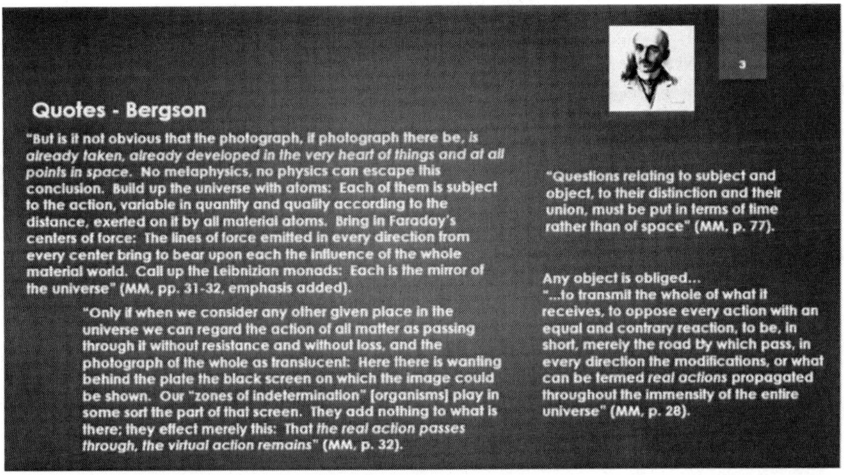

And so, I'd gone over this particular quote from Bergson a number of times -- and for context we should set it up (see above). Bergson is asking: “How do we see the external world? How do we see the coffee cup out there on the table, or the pyramid and ball out there on the table or that coffee cup, how do we see that?” Because there’s no photograph, obviously, of the coffee cup in the brain; and the neuropsychology, shall we say of his day was already advanced enough to clearly know that you are not going to find a photograph or anything like a coffee cup with a stirring spoon inside the brain.

But he went on (here is the quote):

> But is it not obvious that the photograph if photograph there be, is already taken, **already developed in the very heart of things and at all points in space**. No metaphysics, no physics can escape this conclusion. Build up the universe with atoms: Each of them is subject to the action, variable in quantity and quality according to the distance, exerted on it by all material atoms. Bring in Faraday’s centers of force: The lines of force emitted in every direction from every center bring the bear upon each the influence of the whole material world. Call up the Leibnizian monads: Each is the mirror of the universe. (*Matter and Memory*, pp. 31-32, emphasis added)

So, it dawned on me in the seminar with Shaw that Bergson was talking about a hologram, a cosmic hologram. Again, that the head is in the hologram. And when you look at other quotes that Bergson made, if you upgrade Bergson into the holographic understanding -- and remember, holography wasn’t discovered until 1947 by Gabor. Bergson is writing in 1896. And if you upgrade what Bergson probably would’ve said into the current understanding of holography, you come out with the idea that indeed he’s looking at the brain as a reconstructive wave passing through a universal holographic field. And of course, when we get to Bohm, we'll see that Bohm had in 1980, generalized the idea of the universe itself being a holographic field. And we're just going to get a difference as to how the brain actually works in that particular field.

Pryzdia:

Yes. But Steve, can I ask you a question? I'm curious about something: when Pribram was over there at the University of Minnesota and he is having this conversation with Shaw, and Shaw says, hey, the brain is **in** the hologram. What was Pribram's reaction to this? Because he hadn't met Bohm yet.

Steve:

Pribram wasn't there. I think Pribram gave a talk. But I don't remember him remarking or indicating . . . other than him saying (and I am kind of paraphrasing here) . . . it was sort of like he was obviously in awe of Shaw. I should say that. [laughter] So, you could imagine the wild conversation that Shaw had probably had with Pribram, you know. It probably would've been awesome to be a fly on the wall there. But anyhow, going back to that re-casting of Pribram – that the brain is a reconstructive wave, which Bergson was showing . . . you see, this other quote is interesting because he's saying that you can visualize the universe, the universal field, as a sea of "real actions," he called them. You can visualize real actions as waves for example. And he said, any object in this field – and there is a little quote there. Any object is obliged:

> . . . to transmit the whole of what it receives, to oppose every action with an equal and contrary reaction, to be, in short, merely the road by which pass, in every direction the modifications, or what can be termed *real*

> *actions* propagated throughout the immensity of the entire universe. (*Matter and Memory*, p. 28)

But now consider organisms in this field -- an organism (a cat or us) -- we're connected to the field in the sense of we need to be able to act on the field. We're acting systems; we have systems for action. So, in this holographic field, which is information everywhere in the field, we're only interested in the action. The cat, the frog, etc. are interested only in the action which is relevant to the body. So, Bergson goes on and says in the next quote:

> Only if when we consider any other given place in the universe, we can regard the action of all matter as passing through it without resistance and without loss, and the photograph of the whole is translucent: Here there is wanting behind the plate the black screen on which the image could be shown. Our "zones of indetermination" [organisms] play in some sort the part of that screen. They add nothing to what is there; they effect merely this: That *the real action passes through, the virtual action remains.* (*Matter Memory*, p. 32)

So, Bergson succinctly described perception as virtual action as well. So, what you have is the brain acting as a modulated reconstructive wave only responding, picking up in that holographic field that it's embedded in, that information which is useful for its action systems. And then specifying the object or the event.

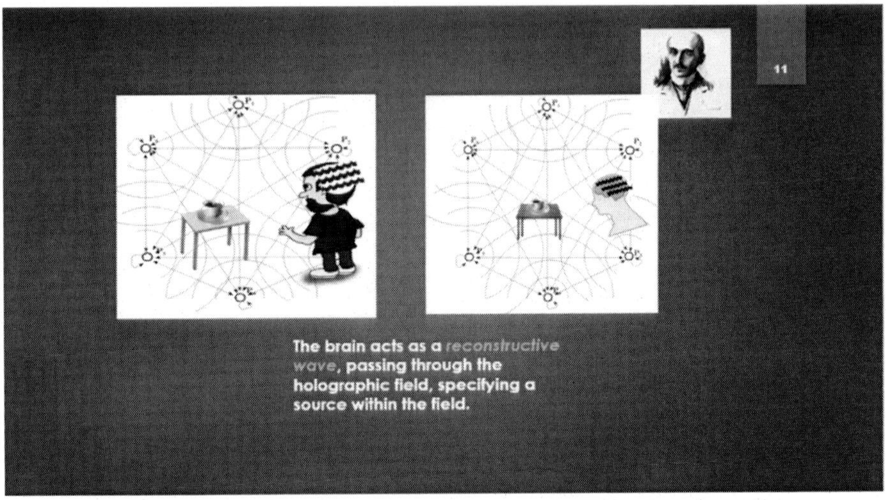

I put a little picture of it here (see above). The best way to look at it . . . there is my holographic field representation, all the lines going everywhere. And the brain is the reconstructive wave, passing through that field specific to a coffee cup being stirred on the table, on the kitchen table, in front of us. However, this his is where you need more than this by far to make Bergson's theory actually go. Because Bergson's theory runs with a different model of time, a different metaphysic -- what I call the "temporal metaphysic" as opposed to the "spatial metaphysic" -- which runs all of science. And you need that metaphysic to understand that other koan -- namely why is there a relationship of subject and object in terms of time? So that's an interesting juncture in the story with me and Shaw. Because I began at that point deciding . . . well, I was definitely going to do my thesis on Bergson and Bergson's model of mind; but I intended to integrate Gibson because in my mind, obviously with Shaw's approval to some degree, J.J. Gibson and his theory of direct perception needed Bergson to work – to actually be comprehensible. Gibson's theory of direct perception is simply not comprehensible to people. He just says, well, the brain, the environment, is "specific to" such and such, and the brain is specific to the environment.

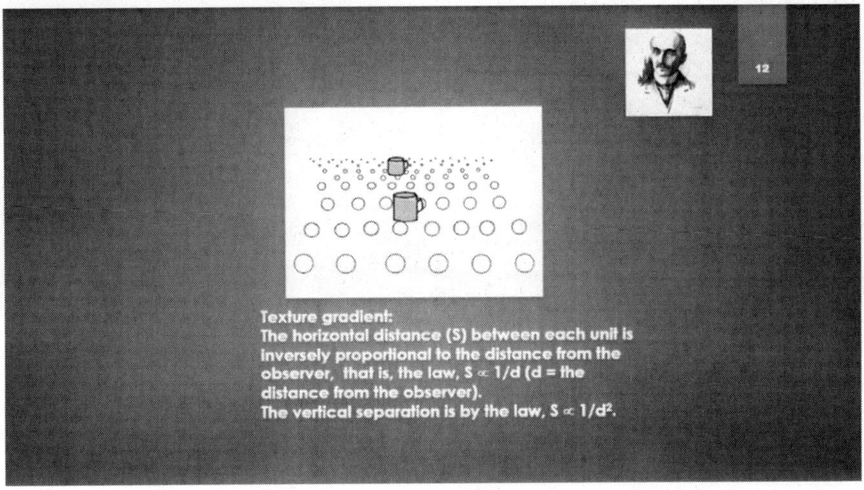

When Gibson is talking about that, he's talking about these kinds of mathematical relationships like the texture gradient (see above) where you'd see this kind of a gradient across, well, your lawn/grass or your gravel driveway or a beach -- they're ubiquitous. And there's this mathematical relationship where the horizontal distance between each of these little texture elements is proportionate to the distance away from the observer. And the vertical distance up and down is proportional to the distance squared from the observer. So, we have this nice law. And if you move the cup back and forth across that gradient there of texture elements, there's a constant ratio of the height of the cup to the number of texture rows that the cup occludes. So, if the forward cup occludes two rows, if it halves in size, moving backwards, it's also obscuring now four rows of the gradient. So, you've got a constant ratio, it's specifying the constancy, the size constancy of the cup. So that's one of the kinds of things that he's looking at.

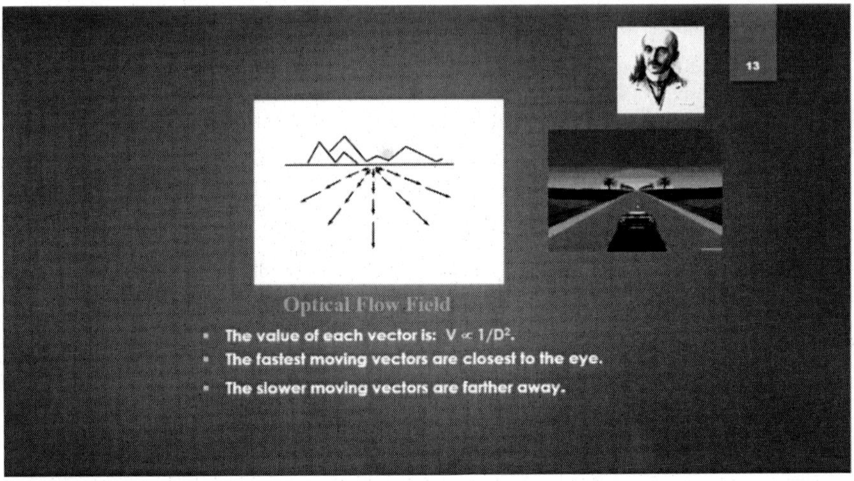

And then of course, you could put the whole field in motion with what he calls an optical flow field -- and then you have these little velocity vectors radiating from a central point as you flow down the road or toward the mountains (see above). And again, the value of each vector, the velocity value is proportional to one over the distance squared from the eye of the observer.

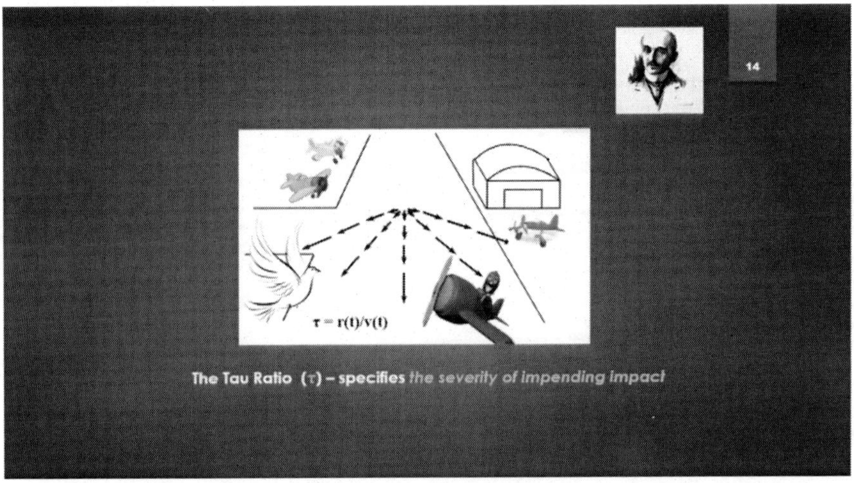

And over these kinds of fields, you'll get interesting ratios like the tau ratio here, which seemed to describe the severity of impact (see above). So that if you're going to make a landing like an airplane or a dove landing on a runway for that matter, you're using the information in that tau ratio to modulate your action systems to keep the impact from being too much – so you can make a nice landing.

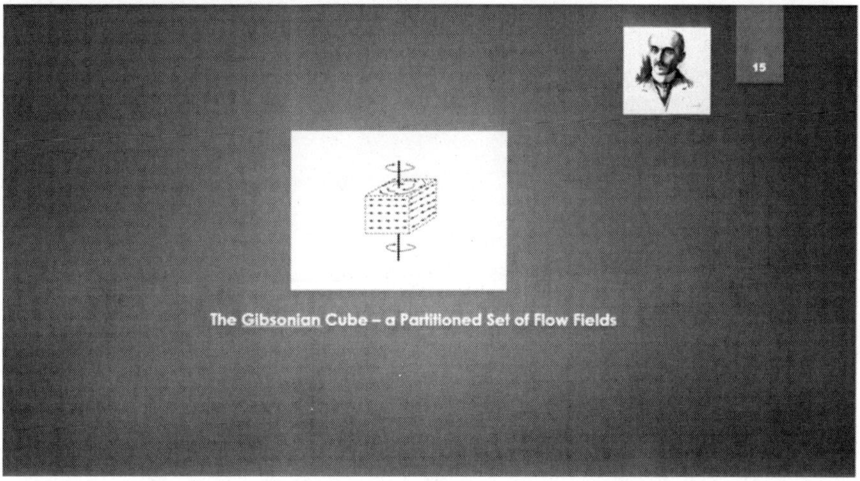

And then it's even worse because then, regarding form, you'll see that these fields are very ubiquitous in form (see above). Here's a nice example of what I call a "Gibsonian Cube." If that were rotating, you'd have a set of those flow fields, you'd have a partition set; you'd have an expanding flow field as the side rotated toward you, a flat flow field for a brief instant -- and then as the side rotated away from you, you have a contracting flow field, you have a radial flow field on the top. And in reality, what are considered static features of that cube, those edges and vertices are now seen for what they are. They're only sharp junctures, discontinuities in those flows. So over time, when you remove the static notion of that cube, it's a dynamically rotating cube, its features are purely functions of these flows and exist very ephemerally as discontinuities in flows.

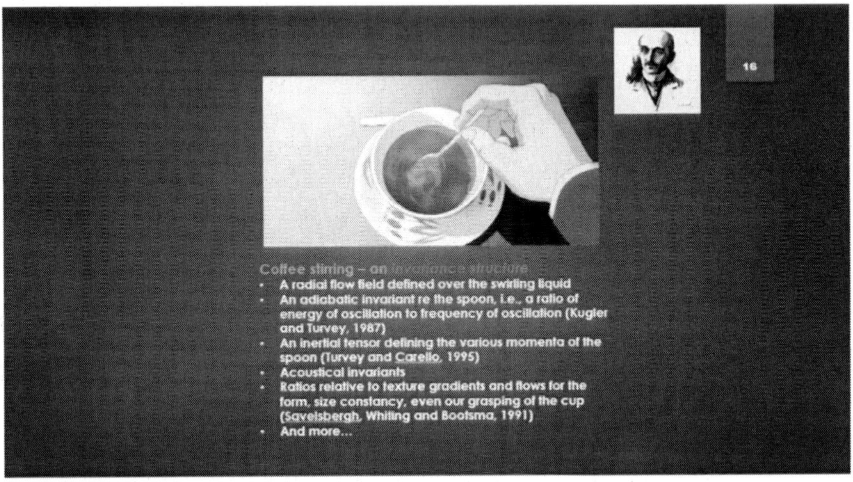

In fact (see above), you take all these kinds of laws in a composite, shall we say, for an event like coffee stirring, you've got a ton of these -- what are called **"invariance laws."** You have a radial flow field defined over the swirling liquid. You've got what's called an adiabatic invariant -- it's a ratio of the energy of oscillation to the frequency of oscillation. You've got inertial tensors defining the various momenta of the spoon. There are acoustical invariants. There are those ratios relative to the texture gradients and flows for the form and size constancy of the cup -- even grasping the cup, because there's something called grasping tau, just like there's landing tau, et cetera. So, to bottom line where Gibson fits in, is that it's all this information that is modulating the brain as a reconstructive wave. It's these invariance laws. **It's the invariance structure of these events that's modulating the brain as a reconstructive wave and specifying then the event externally. But always we're talking about over time; this is a dynamic event.** And that's where Shaw and I started to diverge a bit.

Pryzdia:

What do you mean by that, Steve? How did you diverge?

Steve:

By that you mean why did we start to diverge?

Pryzdia:

Yes.

Steve:

Well, Bergson's model is a four-dimensional model of mind, of being. OK? I'll just say that. It's a very four-dimensional model. The problem is then you ask instantly, well, there's this physics theory called relativity, which is supposedly the authority on time. So how would you reconcile relativity to Bergson; and are they compatible? And, of course, there's this notion that relativity describes what's called a space time block. That is, the whole universe is laid out in time – past, present, and future -- in one giant block. And of course, the question there becomes, well, how do you get any motion then in that block? Or how do we experience the coffee cup being stirred? You end up with weird theories like fields of consciousness passing over the 4-D block, trying to register motion. And, of course, that becomes an infinite regress very quickly. But I found a chapter by Milic Capek in a book, I think it was *Voices of Time* by J.T. Fraser (1966). And Capek was trying to reconcile relativity to "becoming." In fact, I should have added that Bergson was very much a theory of becoming. That is, that the universe evolves, flows in time, and it's not the static 4-D block of relativity. And Capek was looking at the mathematics of the Minkowski schema; and he showed that, well, there's a tri-partition of events via this mathematics where if you visualize it in terms of light cones, all the events proceeding from a point in the light cone are of course in the future, and the light cone behind the event – that's possible causal events in the past, and then outside the light cone, all the events are relativized.

But you still have the notion of becoming, and that from any point, there's a causal chain of events that are within the light cone that are all causal. In other words, he was reserving becoming. And I thought that was cool; and I wrote up a paper on this and I gave it to Shaw; and Shaw liked it. And it just so happened that Gibson was visiting the University of Minnesota at the time. This is like 1975 (and you see how I'm making progress on my PhD very slowly) [laughter] -- because I was writing this theoretical thesis of which only one other had been done in the psychology department back then -- well maybe a couple others. But Gibson read it. And the paper came back to me, and he had scribbled something on the top and basically, to paraphrase him, he said: "Well it's really good to look at big problems, but you probably should look at something else." And I said, well, he obviously didn't like this, but I wasn't sure why because he didn't say anything. Well, that day he was giving his talk at the University of Minnesota, and out of the blue he says this, and I'm, let me find a little quote here. He says (see below):

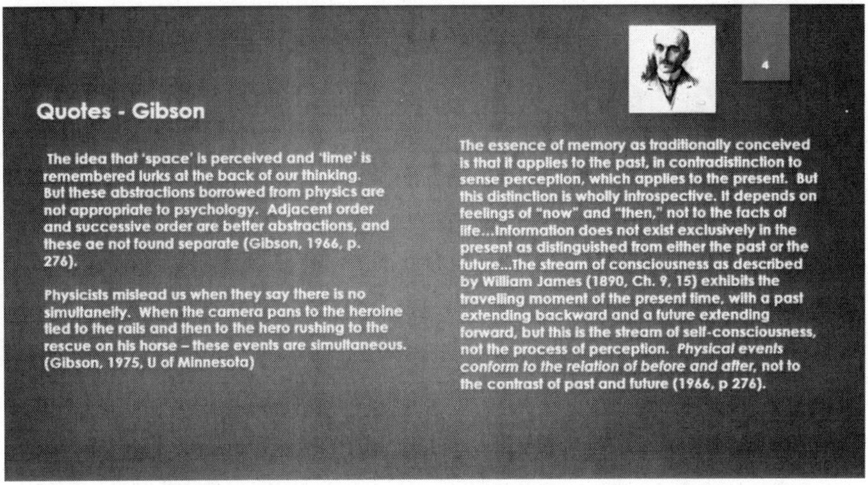

> Physicists mislead us when they say there is no simultaneity. When the camera pans to the heroin tied to the rails, and then to the hero rushing to the rescue on his horse, these events are simultaneous. (Gibson, 1975, University of Minnesota)

And I'm sure that his words just passed by everyone there because it came out of the blue. Unless you'd written the paper I had just written and had him read it, you didn't know why he was saying this. But it struck me like a thunderbolt because I realized that he was siding with Bergson. And why do I say that? Well, in 1922, Bergson had become aware of Langevin's "twin paradox" in relativity where the one twin leaves the earth at high speed and comes back, and he's aged less than his brother who was left on the earth.

And Langevin said this was an implication of relativity. And Bergson was shocked. Bergson felt that this was destroying the logical fabric, the logical consistency, of the theory of special relativity. He was absolutely right. And he had ultimately written a book entitled *Duration and Simultaneity* (1922), where he had taken apart the equations of relativity in complete detail -- all the Lorenz equations; he reconstructed everything; his purpose being to show that there's no way that the twin paradox can be true. That is, time is an invariant flow for both observers; both twins experience the same time flow; and one doesn't age more than the other. There are various reasons for this. I don't want to go into here; but that's kind of where I say I started to diverge from Shaw. Because when you look further at Gibson's statements on time . . . I'm sorry, I should go back one more time to finish that little thought off about Gibson's quote where he says that "physicists mislead us when they say that there is no simultaneity" -- what he's talking about there, when we have the heroine tied to the rails kicking her legs and struggling, and the hero rushing and horses galloping . . . these are simultaneous flows of events. I would call them simultaneous causal flows. And they cannot be relativized. And this is one of the great sins of relativity. They take everything down to a point. All the mathematics is always taken as an instant, as a cross section, an instantaneous cross section of these flows. But the flows cannot be relativized. And Bergson was talking about this in his book in 1922, *Duration and Simultaneity*. To give an example of one these flows (see below):

I like to think of a rose that is blooming (one needs to visualize this) – an organic motion; and let's make it a very wide rose, like 10 feet across. And you have two opposite petals that strike two points, 10 feet across, two points separated 10 feet across; they strike at exactly the same time. Now in relativity, the physicist could fly by in a rocket ship, declare that the observer stationary watching the rose would declare that the two pedals had hit simultaneously on each side of the rose at two points; and he could say that actually you are in motion and your clocks are not in sync. And it was not simultaneous. But the problem is that that is absurd -- because you cannot relativize the organic simultaneous flow and growth of the rose -- no more than Einstein's example of the lightning bolts, where in reality, the lightning bolts are part of an organic evolving simultaneous set of flows of a storm system. And yes, if you take an instantaneous cross section, you can say it's relative. That is, the two lightning bolts either did or did not strike simultaneously depending on the motions of the observer, but in the larger context of the simultaneous flows of the storm or of the rose, et cetera, that's absurd. They can't be relativized. Relativity is just plain wrong. This is what Gibson was arguing and this is what Bergson was arguing.

Then it leads to a more profound problem with the nature of time, because what Gibson was implicitly arguing for in his works was what I call Bergson's temporal metaphysic. In the classic metaphysic of space, the spatial metaphysic, time is reduced to a series of instants. There's a present instant then that falls instantly into the past. In the event flow, there's a present instant that falls into the past – which is the symbol of non-existence. Then you have the next present that falls away as the next present comes -- instant after instant after instant. And the past is simply lost into non-existence. And this comes about from a spatial treatment of time where we are looking at time as a series of instants.

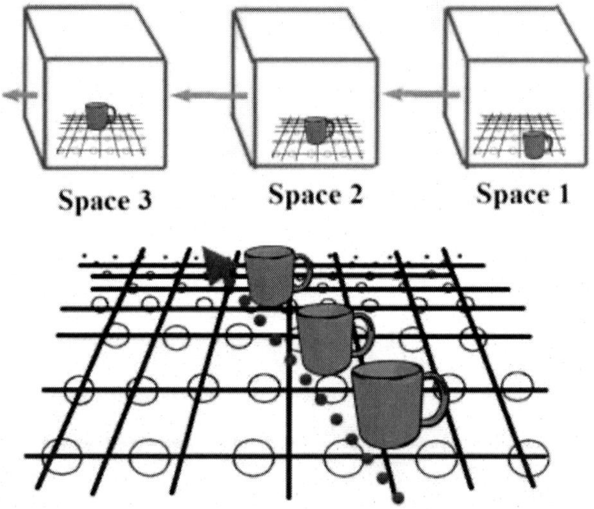

We look at a cup going across a little gradient -- a cup going across a surface -- where that surface is conceived as a continuum of points (see above). And the path made by the cup from point A to point B is a line which then we can divide infinitely into a series of points. And, of course, this divisible line then, between each pair of points, we now have to explain the motion -- if that's our explanation of motion -- between each pair of points, we have to put another line. Again, it's infinitely divisible with an infinite number of points. And we quickly end up with an absurdity. We're trying to reconstruct motion as a series of immobilities and it's obviously an infinite regress.

Bergson argued in the temporal metaphysic that we have to treat motion as indivisible. We cannot divide it into a series of instants. And he pointed out that the whole of all four of Zeno's paradoxes rested on this problem of taking the motion in terms of a space, a line infinitely divisible. But in reality, the motion is indivisible. And not only should we look at the motion as indivisible or the time transformation of the universal field as indivisible, but that we should see it in terms of a melody -- where each instant or note in the melody is the reflection of all the proceeding notes; and each note interpenetrates or permeates the next note create -- creating an organic continuity. Thus, back to the organic continuity of flow. And this should be the model of time.

And what this does is allow for the brain, that reconstructive wave, to specify events, like the coffee cup being stirred, that are actually in the past -- as we would call it in the classic metaphysic – because we are looking at events as restoring that coffee that have long since come and gone or the fly buzzes by the coffee with all his wings beating. But it's long since come and gone. In standard psychological theory, we now try to begin to figure out just how it is that the brain can store all of these little instants and build it up into a whole time-structure so that we can perceive the event. See, suddenly you begin to see why Bergson said whether the brain stores memory is critical to consciousness. Because our consciousness is four-dimensionally extended. We see time extended events; we see the coffee stirring; and there is no theory that is coherent at all. It could all be driven into absurdity. For example, if you're storing let's say each slice of the coffee stirring -- like a series of photographs -- of a slice after slice of the coffee stirring as though they're all laid out in the brain like on a desktop, okay, you've got a bunch of static images. How do you reintroduce the motion again? By some sort of internal scanner? Well then how does the scanner register a motion? The whole thing can be driven into absurdity in many different ways. So, I'll simply state that I laid this out in a paper called "Is Experience Stored?" -- going through a current memory model. There is no coherent model whatsoever of how we see this coffee being stirred -- because it is a massive problem of memory. But in the Bergsonian temporal metaphysic with a time transformation of the universal field that is indivisible and like a melody, each note permeating the next, each instant permeating the next, not divisible -- then the universal field itself carries an elementary property of memory (stealing a concept from William James) -- it's a "primary memory"; it truly is the primary memory. And so, the brain can specify – as a reconstructive wave -- events in the past -- because they really haven't fallen into non-existence in the sense of the classic metaphysic. It is still specifying or resonating to a time period of the transformation of the coffee being stirred.

And it gets a little bit more interesting yet too -- because it's all being specified at a scale. The brain is imposing a scale of time on this field. When the fly buzzes by the coffee cup, you're seeing the fly as a blur, a couple hundred wing beats per second being seen as a blur. But that's the function of we'll say the chemical velocities underlying the brain's neuro processes. Chemical velocity has a very precise definition in the Arrhenius equation. And it can be changed simply by raising the temperature; it can be increased by raising the temperature, or it can be increased by some form of catalyst that makes more energy available to make the chemical reaction occur more quickly. And if you do that little exercise -- and Bergson had clearly thought about this -- if you increase the chemical velocities underlying this resonant wave that the brain is creating, you're changing this scale of time that is being specified.

So, if you increase these chemical velocities, the fly starts flapping his wings slowly like a heron, and then keep on increasing even more. And then pretty soon the fly is stable, it is motionless. Okay? And so, you're getting a couple interesting things going on there right away. Number one is back to the virtual action principle -- perception is display of possible action or virtual action. Well, because all this information is being relayed in the brain to the action systems via feedback from the visual system and back, you are now seeing constantly how you can act on that fly. You're seeing in the original fly, the normal scale of time, the buzzing fly, you could reach out very quickly and maybe grab the fly. By the time the fly is a stable motionless fly with his wings not even moving, you're seeing how you can reach out very leisurely and pick the fly out of the air by the wing tip.

That is the implication of the notion that perception is virtual action. It's also the implication of Gibson's "affordances." The Gibson people just don't realize this. Affordances in the Gibson terms is just another way for saying that perception is virtual action. You're seeing what action the environment affords the organism. Secondly, the interesting thing is you're now seeing why subject and object needs to be seen/understood in terms of time, not space. Because if we invert the process, we start from the body being in this holographic field. But at a scale of time that is . . . well, the scale of time is such . . . how can I put this . . . the body and the fly in this field are, they're not separated spatially. As

Bergson would say, all the objects have . . . we see precise outlines in our perception at this normal scale. But at other scales the objects blend into each other in this holographic field.

So, you can visualize a start where the brain is specifying at a very high energy state; say you're watching the fly as a cloud of electrons -- and you're also a cloud of electrons -- if you look at your hands, et cetera. But as you start lowering the energy state, the chemical velocities as I called it, the fly starts to transition from an electron cloud to now a crystalline vibrating but stationary-like fly to now a fly that has started to slowly move his wings, and now he starts to flap his wings, sort of like a heron. And now you're back ultimately with enough change of energy state to the buzzing being wings of blur of normal scale. So, in that transformation we just described, subject is differentiating from object in terms of time, but not of space. You've never broken the essential unity of the two: your body and the little fly in that field. But what you've done is impose successively greater scales on time to where you start to differentiate yourself as a being in a body from that fly as a being in a body. Hence, subject differentiates from objects in terms of time, not space.

So, just getting back to where I diverged from Shaw . . . Gibson was quite opposed to the classic metaphysic. You have quotes like this:

> The idea that 'space' is perceived and 'time' is remembered lurks at the back of our thinking. But these abstractions borrowed from physics are not appropriate to psychology. Adjacent order and successive order are better abstractions, and these are not found separate. (Gibson, 1966, p. 276)

So that's quite a prohibition for psychology. It's saying you can't use physics' abstractions of space and time. And then utilizing another quote -- and I'm referring to his 1966 book, *The Senses Considered as Perceptual Systems*:

> The essence of memory as traditionally conceived is that it applies to the past in contradistinction to sense perception, which applies to the present. But this distinction is wholly introspective. It depends on the feelings of 'now' and 'then,' not to the facts of life . . . information does not exist exclusively in the present. (1966, p. 276)

Now remember that present moment in the classic metaphysic, that instantaneous present -- he's saying there is no such thing as this instantaneous present. And so, he says,

> . . . information does not exist exclusively in the present as distinguished from either the past or the future . . . The stream of consciousness described as by William James (1890, Ch. 9, p.15) exhibits the travelling moment of the present time, with a past extending backward and a future extending forward, but this is the stream of self-consciousness, not the process of perception. Physical events conform to the relation of before and after, not to the contrast of past and future. (1966, p. 276)

So, what he is saying there is, "Yeah, I don't care about your classic metaphysic with its line of instants, one being present, the rest in the future, and the others in the past, and then we move to the next present and the former present drops away into the past. There's no such thing." We are not dealing in terms of this particular metaphysic with its present instant, its past, and its future instants. That's why we're talking of "before" and "after" relationships. But "before" and "after" has nothing to do with a "present" and a "future" – in the sense of the classic metaphysic. So, when I say I diverged from Shaw . . . what we've just gone through is . . . you see how Bergson's temporal metaphysic is required to make his model, this reconstructive wave model, work. And similarly, Gibson's model . . . in fact I neglected to say with regard to Gibson's "direct perception," he's saying that the brain is "specific" to the environment, just like the reconstructive wave is specific to an object wave: the cup or the pyramid and ball. But no one's been able to understand that because the Gibsonites have never brought in Bergson's holographic model. And secondly, they haven't understood this prohibition that Gibson made: we can't bring in the abstractions borrowed from physics -- they're not appropriate to psychology. But what Shaw did ultimately, because he loved physics, he has a brilliant physics mind, is try to model ecological psychology on quantum mechanics. And Turvey, another brilliant collaborator with Shaw, was trying to model direct perception on quantum mechanics also. Shaw was trying to model it on the wave equation actually. And you see, Gibson would say, "No way, you're barking up the wrong tree."

And I just saw another article coming out where the author is toying with using that four-dimensional space-time block and the implications that has for discussing memory in Gibson. Because remember, Gibson made other statements. I'm not sure I have them though . . . Well, I don't have the illustration here where Gibson literally trashes the whole idea that memories are stored in the brain – that it is a stultifying notion. But here's another quote from Gibson (see below):

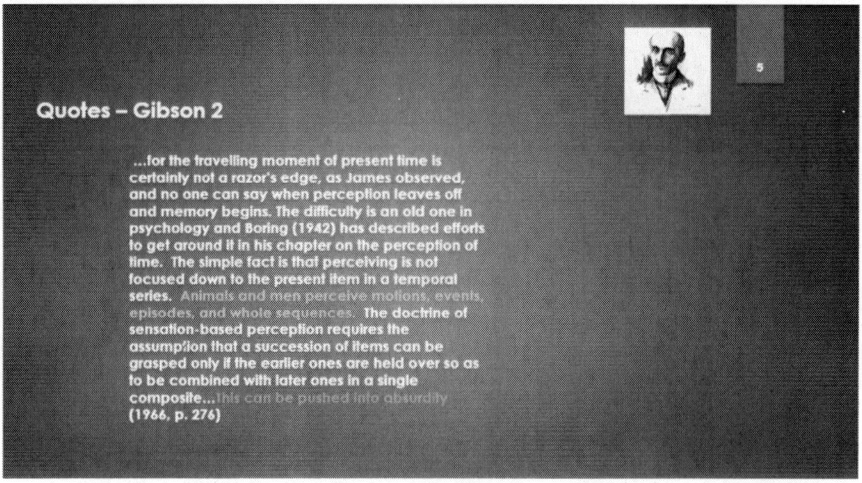

> . . . for the travelling moment of present time is certainly not a razor's edge, as James observed, and no one can say when perception leaves off and memory begins.

Put a parenthesis there for a second . . .

In reality, in the Bergson model, perception is **always** a memory. You're not trying to find this hypothetical combining of perception and memory. Perception is always a memory -- because it's the specification of past transformation, indivisible transformation of the field. But going on with the Gibson quote from above:

> The difficulty is an old one in psychology and Boring (1942) has described efforts to get around it in his chapter on the perception of time. The simple fact is that perceiving is not focused down to the present item in a temporal series. **Animals and men perceive motions, events, episodes, and whole sequences**."

In other words, you are seeing time extended dynamically transforming events. You need the primary memory of the universal field as it transforms, as specifying this. Gibson goes on:

> The doctrine of sensation-based perception requires the assumption that a succession of items can be grasped only if the earlier ones are held over so as to be combined with later ones in a single composite. . . . **This can be pushed into absurdity** (*The Senses Considered as Perceptual Systems*. 1966, p. 276)

And that was my example of taking slices of the coffee stirring -- succession of items -- and placing them one after the other on a desktop in the brain, so to speak. And then saying somehow, you're going to get motion back out of that static structure you've placed in the brain. Not going to happen. And it can be pushed into absurdity because now you're led to scanners, scanning it, et cetera.

So that is where we diverged; ecological psychology got taken very much into physics and modeling after physics. This is not to say that mathematics shouldn't be employed or can't be employed, but it's got to be put in the context of the temporal metaphysic.

Pryzdia:

The biggest challenge, Steve, is this temporal metaphysic. I mean it's challenging for people working in physics, psychology, consciousness studies today -- that that's the thing that . . .

Steve:

Right. Yes. And as we were talking, before we started recording -- about my looking at the review of this computational neuroscience conference, and noting that it's just a massive endeavor -- it is a juggernaut. But the problem is that the whole of computational neuroscience and a whole computational metaphor and the whole of AI is based in the classic metaphysic. It's nothing but a focus on an abstract series of states -- state after state after . . . just think of the Turing machine and that tape, that tape divided into discrete little sections, one after the other after the other being moved. There's never more than one instant, one instantaneous instant in the computer metaphor. **There is no possibility of the computer model dealing** with **time extended transforming events.** And this is why I've noted that Penrose, when he talked about **non-computational thought** in his 1996 book, *The Shadows of the Mind*, for which he was roundly hated for -- what was he visualizing?

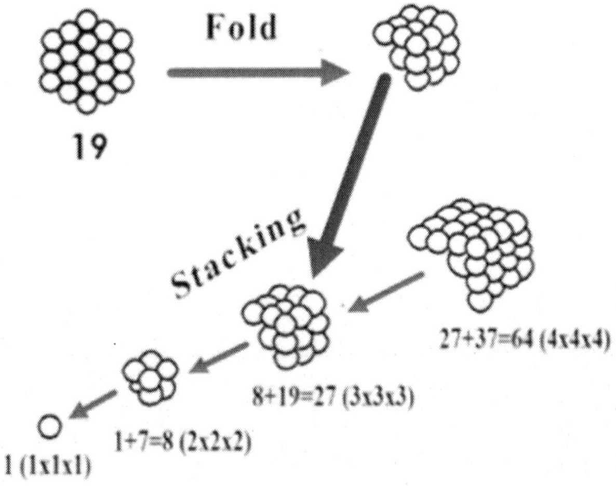

He was visualizing (see above) a hexagonal number, like a 19, that could be folded into a three-sided cube -- the three-sided cube lacking one of the sides. And he's looking at a computation that does not stop. That is, if you add successive hexagonal numbers, you always get a cubicle number. And so, what he does is he takes a cubicle number, starts with, like 1 is a cubicle number, takes the next hexagonal number, a 7, folds it into a three-sided cube, stacks it over the first one, now he's got an 8; then he takes the next hexagonal number, a 19, pulls it into the three-sided cube, stacks it over the previous number, the previous cube, which was an eight, and now we have 27. But his argument is you can see visually that this just keeps going and going and going. You're always going to get a cube. And the problem there is he didn't realize that **you can only see this in an indivisible time.** It's only in an indivisible time flow of perception, whether it's mental or whether it's in the actual concrete perceptual world, that one can see the globality of the transformation and the invariance being preserved. So that's why it's non-computational. It has nothing to do with the computer metaphor of abstract states, which is the epitome of the classic metaphysic. So again, you will find . . . I saw people in ecological psychology . . . they're trying to dump the computer metaphor, but **it doesn't do any good to dump the computer metaphor if you're keeping the classic metaphysic on which the computer metaphor is based** – that's the core of the problem. And it's also the core of the qualia problem. If you notice, we've been describing qualia all throughout this -- or qualities of form, of rotating cubes, of buzzing flies versus heron like flies versus crystalline like flies - at various scales of time. Because when you remove the classic metaphysic . . . when you go to the temporal metaphysic, you are specifying qualitative motion in a qualitative field at a scale of time. And the hard problem is arising from this classic metaphysic and also frankly from the problem that they don't understand that you're trying to explain the origin of the image of the external world in the first place.

Pryzdia:

That's true.

Steve:

They think they're just trying to explain qualia. But **form is qualia**. Everything in the external world image, our image of that coffee cup and the table, et cetera, is qualia. **The whole** image **is qualia; you are trying to explain how the image is perceived.**

Pryzdia:

Form is qualia; that's what people struggle with.

Steve:

Right. So that's kind of bringing us to Bohm.

Pryzdia:

Yes. This is a good segue into Bohm. I don't know if you want to go into Bohm now or wait for another time. Because we've been going . . .

Steve:

Maybe we should wait here. Yeah.

Pryzdia:

But this is a very good segue. To set us up for next time -- let's talk about Bohm and his relationship to the temporal metaphysic versus the classical metaphysic. He's very unique. As you know, he's so close to Bergson. But at the same time . . . we have to for instance talk about . . . I think it'd be a very good idea to talk about the difference between direct and indirect perception -- or direct and indirect realism. Because you're right on when you say Bohm is an indirect realist. I think that's a good place to start. Because that ties into the whole temporal metaphysic thing as well. So let's start there. Let's get into Bohm next time. That's a good plan.

Steve:

I think that's a good idea.

Pryzdia:

Is there anything you wanted to say here before we stop though, because this was a good introduction to the whole Gibson-Bergson integration. Do you want to say anything else?

Steve:

A thought just flashed by me and I just lost it. Nope. I will just say no.

Pryzdia:

We're done. Alright. I'll stop recording here.

Talk 2: On Bergson, Pribram and Bohm

Pryzdia:

Okay Steve, let's dive now into Pribram and Bohm. Take it away.

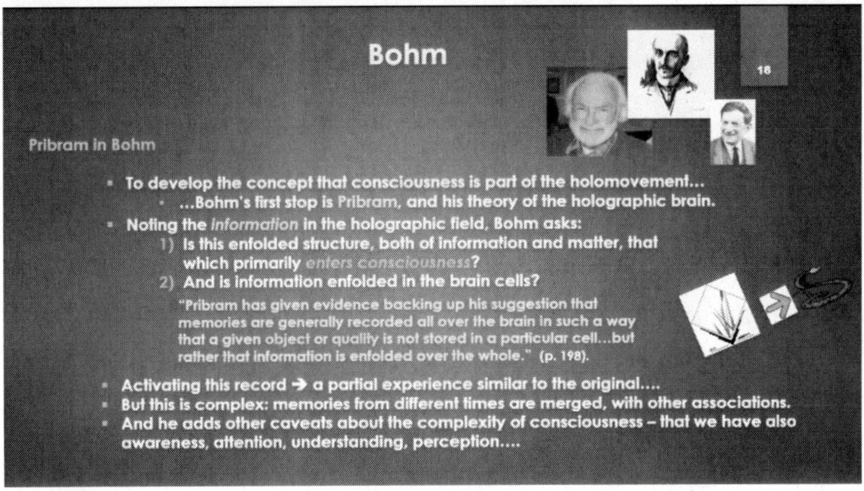

Steve:

Sounds good. I made a few slides here to help us flow through this thing a little bit more quickly, but obviously the key thing in Bohm is that he's got consciousness as part of the holomovement, and his first stop in discussing this . . . if we look at his book . . . this is all relative to the *Wholeness and the Implicate Order* book of 1980 . . . his first stop is Pribram and Pribram's holographic theory of the brain. And noting that information is in the holographic field, Bohm asks: Is this enfolded structure of both of information and matter that which primarily enters consciousness? And is information enfolded in the brain cells? And then he quickly goes to Pribram here.

> Pribram has given evidence backing up his suggestion that memories are generally recorded all over the brain in such a way that a given object or quality is not stored in a particular cell . . . but rather that information is enfolded over the whole. (p. 198)

So obviously, and implicitly if not explicitly, Bohm is endorsing Pribram's view that memory/experience is stored in the brain. And I believe we discussed that in Bergson's model, experience is not occurring solely in the brain and therefore cannot be stored there. For Bergson the brain is specific to the external environment. The cup out there on the kitchen table is right where it says it is, and that experience is not in the brain. It's being specified out there as part of the past transformation of the field. So right away we've got a problem going with both Bohm and Pribram in that with both of them we are adhering to the "storage in the brain model" – storage of experience in the brain. And then he notes: activating this record that is stored in the brain, you get a partial experience similar to the original. But he notes: this is complex: memories from different times are merged with other associations. And he adds other caveats about the complexity of consciousness -- that we also have awareness, attention, understanding, perception -- all of which is true.

This entire notion of the partial experience similar to the original being restored is part of a discussion that I've given on the concept of redintegration where a present experience has certain pattern or invariance laws that cause the brain to go into a resonate state that becomes resonant, shall we say, with a past experience in the 4-D extent of being. But that of course is not Bohm's framework here. Actually, I'm symbolizing "redintegration" here in this illustration. In other words, a pattern, a rustle in the grass redintegrates past experience (i.e., my seeing a snake crawl through the grass and me being frightened) -- redintegration. That's what they're really referring to here, but quite a different model between Bergson and Bohm.

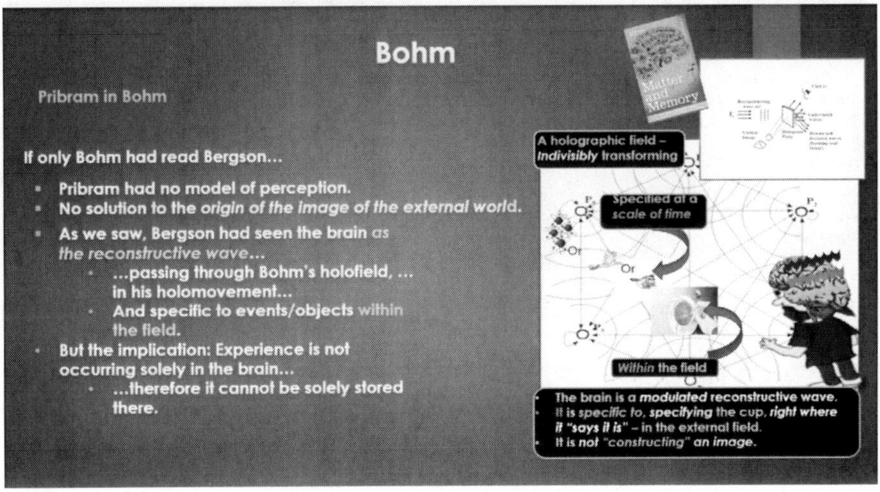

Steve:
So, this is my complaint: if only Bohm had read Bergson. Pribram had no model of perception – that is, he had no solution to the **origin of the image of the external world**. As pictured here (see above), and as we discussed previously, he had nothing like this -- where the **brain is taken as a reconstructive wave** passing through the holographic field just as the reconstructive wave passes through the hologram plate. Here, the brain is the reconstructive wave passing through the holographic field and is specific to a virtual image out there in the field right where it says it is -- the coffee cup and stirring spoon with a buzzing fly going by -- namely a certain scale of time being specified to -- a buzzing fly as opposed to a heron-like fly, or a cloud of electrons. So, as we saw, that's Bergson's model. But the implication as I noted: experience is not occurring solely within the brain -- therefore it cannot be solely stored there.

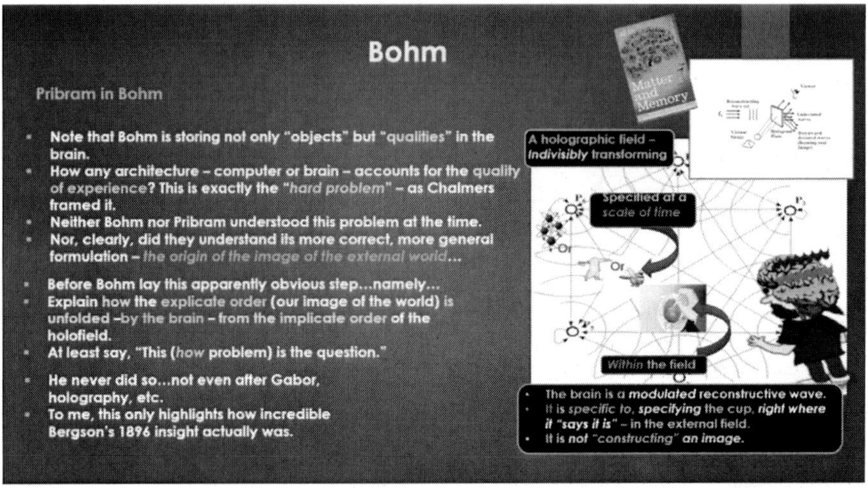

Note that Bohm is storing not only "objects" but "qualities" in the brain. How any architecture -- computer or brain -- accounts for the quality of experience? This is exactly the "hard problem" – as Chalmers framed it. But neither Bohm nor Pribram, shall we say, are really aware of the problem at the time -- it just hadn't really filtered into people's consciousness in . . .

Pryzdia:
It didn't really exist yet.

Steve:
Yeah -- 1972 Pribram publishes *Languages of the Brain*, 1980 for Bohm with *Wholeness and the Implicate Order*. Obviously then they didn't get the understanding, as I would put it, that the more general problem is how you explain **the origin of the image of the external world**. That is, there is nothing in the external image that is not quality -- including all the forms, the dynamically changing form. So, as I noted, before Bohm lay this apparently obvious step . . . namely . . . explain **how** the **explicate order** (our image of the world – the coffee cup out there, the fly going by) **is unfolded -- by the brain -- from the implicate order** of the holofield. At

least say, "This (this **how** problem) is the question." But he never does . . . not even after Gabor, holography, et cetera -- which he had the benefit of knowing about -- which to me, by the way, just highlights how incredible Bergson's 1896 insight actually was.

Pryzdia:
Absolutely, yes.

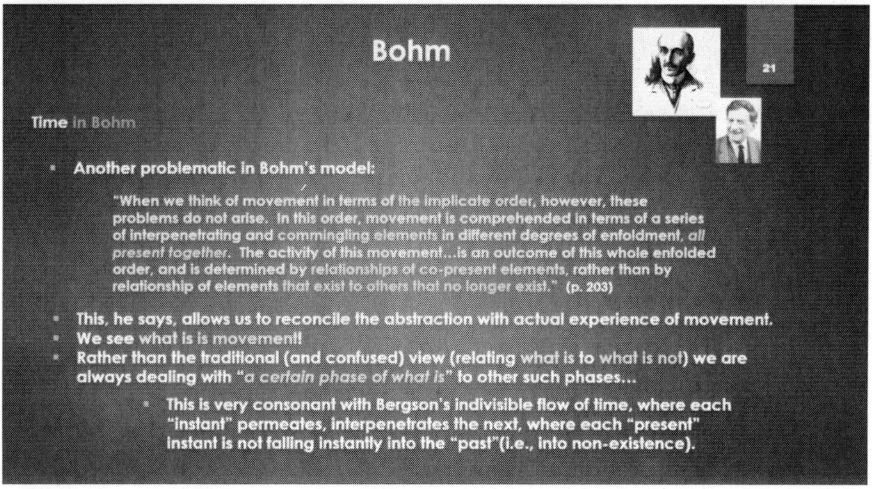

Steve:
Anyhow, that was kind of the major complaint I have with Bohm. If he just would have looked at Bergson and said, boy, yeah -- it's a holographic field and the brain is essentially the reconstructive wave -- that would be how the explicate order is unfolded. That would've been cool. Well, another thing in Bohm that's difficult is time. Here's a quote:

> When we think of movement in terms of the implicate order, however, these problems do not arise. In this order, movement is comprehended in terms of a series of interpenetrating and commingling elements and different degrees of unfoldment all present together. The activity of this movement . . . is an outcome of this whole enfolded order, and is determined by relationships of co-present

> elements, rather than by relationship of elements that exist to others that no longer exist. (p. 203)

This he says, allows us to reconcile the abstraction with the actual experience of movement. We see **what is is movement**! So again, he says rather the traditional (and confused) view (relating **what is** to **what is not**) we are always dealing with "**a certain phase of what is**" to other such phases.

One has to admit that this is very constant with Bergson's indivisible flow of time -- where each "instant" permeates, interpenetrates the next, where each "present" instant is not falling instantly into the "past" (i.e., into non-existence). In other words, you can get this kind of co-present together time-extent of "instants" that Bohm is discussing -- and that you don't have this notion . . . he is sort of denying that you actually have instants that fall into the past, into non-existence -- that is . . . take that last phrase from above . . . "that exists to others that no longer exist." That is, in the standard model of time (the abstract space, the four-dimensional space of the classic metaphysic) as we discussed earlier, the present moment is the only moment that exists. All previous instants have fallen into non-existence, which we symbolize by the past. And in Bergson's model along with Bohm, well . . . they do not fall into non-existence. They're all there in some mode.

Pryzdia:
And right there. Let's comment on that – "in some mode." So, for Bohm, are we saying . . .

Steve:
Well yeah, they're there in a sense that the flow of time, the dynamic transformation of the matter field, as I have been saying, is indivisible -- each moment interpenetrating and

permeating the next. So you've got a four-dimensionally extended transformation. That mode is precisely . . . I don't know how to put it -- what mode to call it . . . other than just the transformational mode of the universe.

Pryzdia:
But it always seems . . .

Steve:
But here's another interesting . . .

Pryzdia:
The issue is: what's the "glue"? I've always felt that for Bohm the glue is the implicate order, whereas for Bergson, I don't know what we call it . . ., memory or we call it . . .

Steve:
Well, the glue is time itself -- the very nature of time is the glue. And that's why we were just discussing before we started recording our talk here, that the motion of time, the motion of the matter field itself is memory. The time transformation of the matter field, the very change of the material universe, **is** memory. There is no way around that.

But interesting quote here from Bohm -- this is the interesting musical metaphor that he uses:

> At a given moment a certain note is being played, but a number of previous notes are still "**reverberating**" in consciousness. Close attention will show that it is the simultaneous presence and activity of all these reverberations that is responsible for direct **and immediately felt sense of movement**, flow and continuity. To hear a set of notes so far apart in time **will destroy altogether this sense of a whole unbroken living movement**. . . (pp. 198-199)

So again, we're very close to Bergson. When Bohm views the motion of time . . . and the best metaphor for it being the notes of a melody . . . where if you take notes as equivalent to "instant," each note/instant interpenetrates, permeates the next. And each present note is a reflection, the state of that present note, the sense by which we feel it, is a reflection of all the other previous notes. You get to some state or to a point in a symphony where a certain chord is played, where its entire significance is relative to that whole preceding period of the symphony. But more simply, in a shorter time frame, "twinkle, twinkle, little star" . . . all you have to do is hold one of the notes longer than normally to destroy the melody. So, you can see he's very close in a sense to Bergson. And yeah, to hear a set of notes so far part in time, it is . . . again in the twinkle twinkle little star phrase/analogy . . . you could destroy the whole recognition of twinkle twinkle little star just by altering the time flow. So, he says how a sequence of notes is **enfolding into consciousness**, and how at any given moment the transformations flowing out of many such enfolded notes **interpenetrate and intermingle** to give rise to an immediate and primary feeling of movement. (p. 199)

So again, it is so close to Bergson, but there's a little bit of a disturbing sub-tone here in that **it is as though you have to have this interpenetration and intermingling to give rise to the movement**. But **it is** the movement – you are experiencing the movement. We don't need to go into that, but I just think there's a little bit of a subtext here that is perhaps going off in the wrong direction.

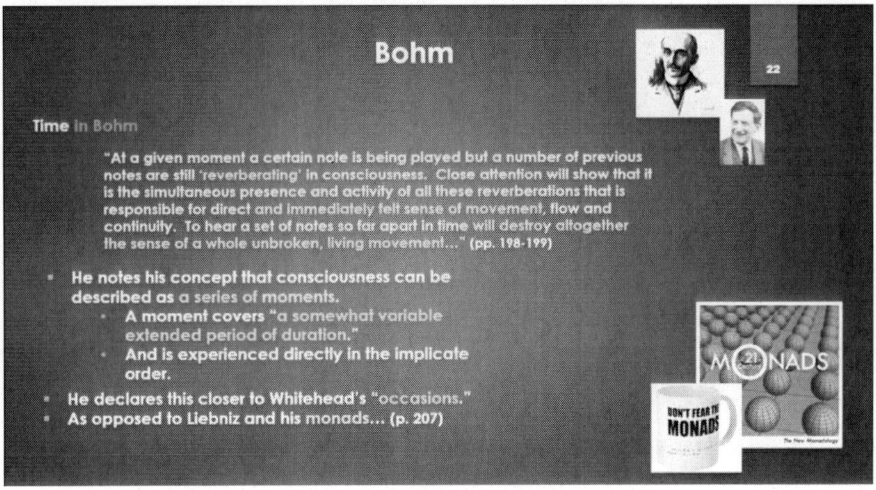

So, he says:

> At a given moment a certain note is being played but a number of previous notes are still 'reverberating' in consciousness. Close attention will show that it is the simultaneous presence and activity of all these reverberations that is responsible for direct and immediately felt sense of movement, flow and continuity. To hear a set of notes so far apart in time will destroy altogether the sense of a whole unbroken, living movement . . . (pp. 198-199)

He notes this concept that consciousness can be described as a **series of moments**. A moment covers "**a somewhat variable extended period of duration**" and is experienced directly in the implicate order. He declares this closer to Whitehead's "**occasions**" as opposed to Leibniz and his **monads**. . . (p. 201).

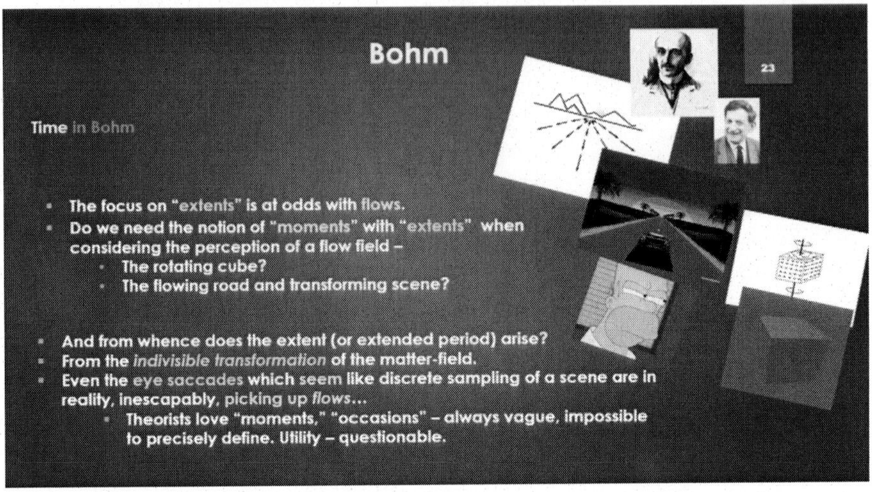

The problem here is this focus on "extents" like "occasions" or "epochs" -- we just really don't need that notion of "**moments**" with "**extents**" when considering say the perception of a flow field or the rotating cube being taken as a partitioned set of flows. Again, that rotating cube . . . as the side rotates towards you, you've got an expanding flow field, as it rotates away, you have a contracting flow field; or the road -- the transforming velocity flow field. How do you partition that into a set of epochs or moments?

Pryzdia:
Now Steve I wanted to bring this up -- and maybe we can build on this later if we have the space and time -- but this kind of does tie into Capek -- and I'm just captivated with his notion of the "pulsations." So, I think what we are saying here is this notion of "moments" with "extents" is kind of like what a guy like Capek would call a "pulsation," right?

Steve:
Right. Yes. And I never got back to you on that I don't think. But yeah, I'm not a fan of that pulsation thing because it comes down to trying to actually define . . . how would you

define a pulsation -- it's going to become very arbitrary, very unprincipled.

Pryzdia:
Let me just say first off, you're the only person I've ever come across who is **not** in favor of that notion of pulsations. All the other Bergson scholars I've read seem to be somehow trying to integrate, let's say Bergson and Whitehead.

Steve:
Yeah.

Pryzdia:
And that to me is a really key piece because in my mind, Steve -- and we can talk about this later maybe -- but it comes back to the notion that Bohm had this notion of a "generative order" from which the pulsation could unfold from. Right? And of course, he's looking at things quantum mechanically.

Steve:
Right.

Pryzdia:
And in my mind, **there is a way to stitch Bergson and Bohm together, but it has to do with viewing Bergson's "duration" as itself a "generative order"** in the sense that Bohm used that term.

Steve:
Right. Well, yeah, we have to remember . . . and we're going to get there . . . I think I have that in a slide . . . He's trying to reconcile the quantum mechanic notion of "jumps" which are pulsations . . .

Pryzdia:
Yeah.

Steve:
. . . into a coherent notion of the motion and flow of time.

Pryzdia:
Yep.

Steve:
And it, it's a good try. And I thought about it for a long time too, until . . . in fact Schrodinger . . . in his 1952 article where he just denies that there are quantum jumps . . .

Pryzdia:
Right? See, that's . . .

Steve:
It is really intellectually freeing.

Pryzdia:
And Bohm was very . . . he wouldn't do that. He stuck to . . .

Steve:
I guess he didn't read Schrodinger's article.

Pryzdia:
You know, in his literature . . . if you read the letters that he wrote to certain individuals that were published after his death (see for example: *David Bohm: Causality and Chance, Letters to Three Women* (2017) edited by Chris Talbot; and *David Bohm's Critique of Modern Physics: Letters to Jeffrey Bub* (2020) edited by Chris Talbot) . . . it appears that he did struggle with Schrodinger. Because as you know, he was always committed to trying in some way to integrate relativity and quantum mechanics, which he never did . . .

Steve:
Right. That's another problem. Yep.

Pryzdia:
Anyway . . .

Steve:
Let's see what we got here. Yeah, so I said [reading from the slide]: And from whence does the extent (or extended period) arise? So again, I am just questioning: well, how do you get the pulsation out of a flow field?

Pryzdia:
Yes.

Steve:
You can sort of almost get a sense of a pulsation from that rotating cube (referring to a graphic on the slide above). But . . .

Pryzdia:
You see, Steve this comes down to the notion that for Bergson, what Bohm would call the explicate order is itself holographic. There's no need for an implicate order. Whereas for Bohm, there is that need for an implicate order from which something could emerge or unfold. It comes down to that. It comes down to the notion that Bergson's holographic field, as you see it, is part of what Bohm would call the explicate order. It seems that for Bergson, there's no need for a generative order. And that is just fascinating to me.

Steve:
Right. It's interesting. And I think if we explored it more, the whole notion of "becoming" in Bergson would be contrasted to how Bohm thinks the universe evolves; it would be an interesting question. But my mind starts to boggle. Let's go on here. Well yeah (referring to the slide above) I noted that even the discreet **eye saccades** which **seem** like discrete sampling of a scene are in reality, as Gibson points out, actually picking up **flows**. The eye isn't picking up snapshots; it's picking up flows -- as a sampling across the flow of that field. As I've said, theorists love the "moments," the

"occasions", the "epochs" -- they're always vague, impossible to precisely define. The utility is . . . I've never seen a use for them.

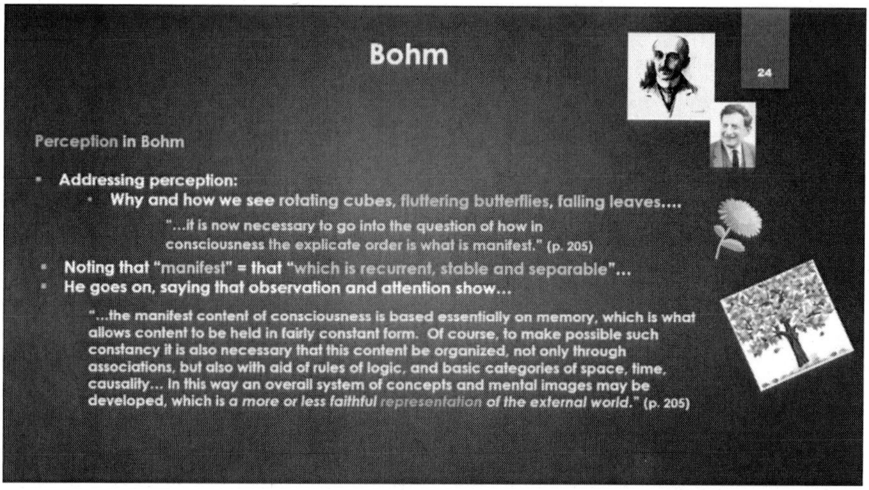

Back to perception in Bohm (in general, not just Pribram): Why and how we see rotating cubes, fluttering butterflies, falling leaves -- he says:

" . . . it is now necessary to go into the question of how in consciousness the explicate order is what is manifest (p.205)." Okay. Which was my complaint. Well, why didn't you just say that it has been unfolded via the reconstructive wave? He notes that "**manifest**" = that "**which is recurrent, stable and inseparable**." He goes on saying that observation and attention show . . .

> . . .the manifest content of consciousness is based essentially on memory, which is what allows content to be held in fairly constant form. Of course, to make possible such constancy it is also necessary that this content be organized, not only through associations, but also with aid of rules of logic, and basic categories of space, time, causality . . . In this way an overall system of concepts and

> mental images may be developed, which is a more or less faithful **representation** of the external world. (p.205)

Now that "more or less faithful representation of the external world" is precisely his declaration of, or vote for, indirect realism. We're not seeing the coffee cup out there on the table where he says it is. We've got an internal image somewhere in a mental space, a perceptual space in the brain, anywhere but in the physical universe, apparently. And that's what we're seeing. And then there are other comments to make here. I hope I get to them because there's . . . in this quote here, there's some confusion or conflation of various sophistications of consciousness, various sophistications of cognitive development, mental levels of consciousness. Obviously, I can perceive the coffee cup out there. Then I can perceive it in such a way that I remember it. I remember having a coffee at the coffee shop with my wife a week before -- and precisely locate that experience in time relative to the time I'm in -- which is a question of explicit memory. And I can even remember and recognize that Juan Valdez in Brazil has had something to do with bringing the coffee to me. But that's all way beyond just the dynamic specification of the brain as a reconstructive wave with a coffee cup out there. So, he's getting way more into cognition than he realizes, and memory.

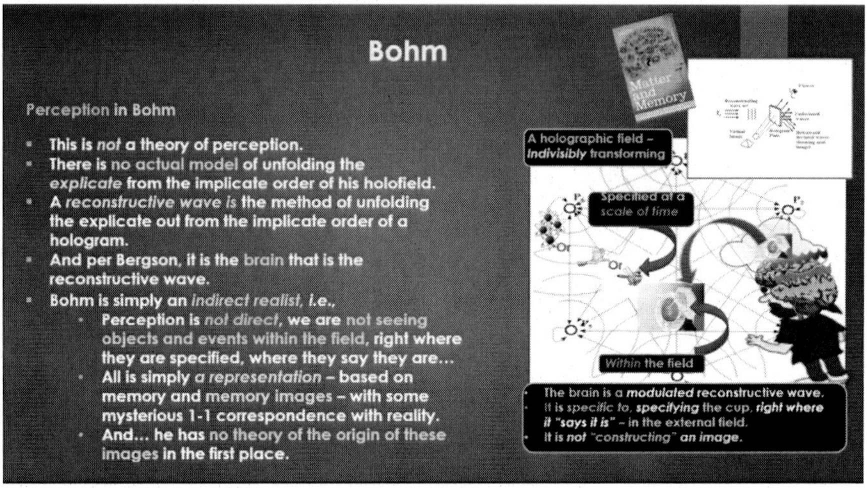

I'm going to say that this is **not** a theory of perception. There's **no actual model** of unfolding the explicate from the implicate order of his holofield. Again, a reconstructive wave is the method of unfolding the explicate out from the implicate order of a hologram (see above). A hologram plate there up in the red box in the upper corner has an interference pattern imprinted on it that looks nothing like the coffee cup or the pyramid and ball. So, it is "implicate" in that sense -- and that is his model. But the object wave, the original object waves or source, are unfolded by the reconstructive wave into their explicate form -- the pyramid and ball and the coffee cup. So again, for Bergson, it is the brain, the dynamics of the brain, that is the reconstructive wave. So, Bohm, as we noted, is simply an **indirect realist**. That is, **perception is not direct**. We are not seeing objects and events within the field, right where they are specified, right where they say they are . . . All is simply **a representation** -- based on memory and **memory images** -- which as we noted in the beginning, he doesn't know how they're stored in the brain in the first place. You can't store your memory images in the brain. There's no coffee cup in the brain; there's no pyramid and ball in the brain, et cetera.

But again, you've got some mysterious one-to-one correspondence with the reality – a magical mystical correspondence.

Pryzdia:
And I just have to say this is because he aligned himself with Pribram -- which he didn't need to do. He didn't need to do that. But somehow . . . they became friends, they got close and he went with . . . he would say to people . . . anytime someone asked him about perception, he would say "Go talk to Pribram."

Steve:
Right. Well, obviously Pribram was the supposed authority on perception and memory in the brain. And so, he was being respectively respectful of Pribram and Pribram's subject matter. Just like Bergson tried to be respectful of Einstein and the physicist.

This is Bohm. We have the indirect representation somewhere in some mystical space, a cloud somewhere, a perceptional space as Joscha Bach likes to call it. And that is a nice one-to-one correspondence with reality.

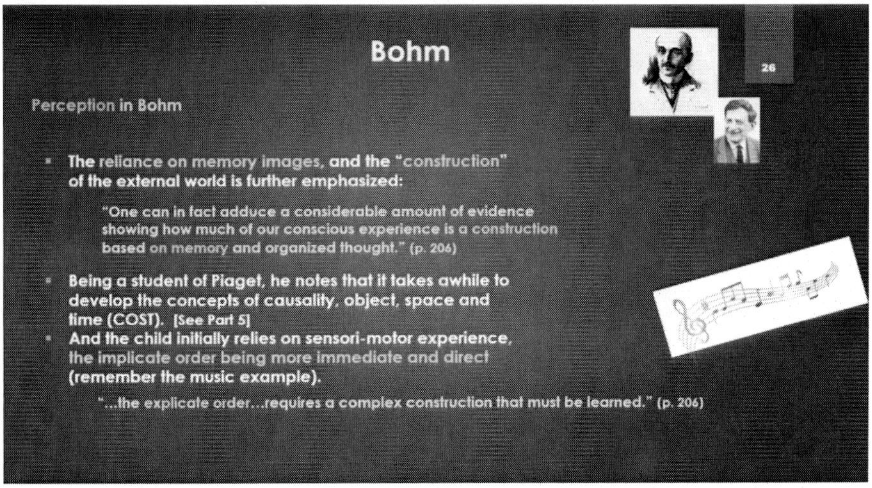

So the reliance on memory images, and "construction" of the external world is further emphasized here:

> One can in fact adduce a considerable amount of evidence showing how much of our conscious experience is a construction based on memory and organized thought. (p. 206)

Well, that's true. Being a student of Piaget, Bohm notes that it takes a while to develop the concepts of causality, object, space, and time. I like to summarize it as COST. And the child initially relies on sensory-motor experience, the implicate order being more immediate and direct (remember the music example). So all this is true, but again, you're talking, as I noted . . . in the child's case of cognitive development . . . in the development of those concepts which take a couple of years (causality, object space, and time) . . . you are now developing how you can see the coffee cup as an experience that's related to an experience that you had a week earlier drinking coffee with your wife in a your favorite coffee shop. You can locate an object or an event in the past explicitly. And that's now part of your perception, but it's not the fundamental problem of how that cup coffee cup on my table being stirred happily is specified in the first place. He just gets things confused here in terms of the basics versus the much more developmental stage along the way, the more sophisticated development. He notes " . . . the explicate order . . . requires a complex construction that must be learned (p. 206)".

Well, no! The explicate order is the coffee cup on the table out there being stirred, being specified by the brain as a reconstructive wave. It doesn't require complex construction to be learned. Only if you're taking that event as a nested event specifically understood as related to something in the past or some other conceptual framework as you're putting it in. Do you see where I'm going with that?

Pryzdia:
Yes. It is a representation for him. That's what it is.

Steve:
Right. And so he is conflating a whole lot of things here. Thus (referring to the slide below), my comment:

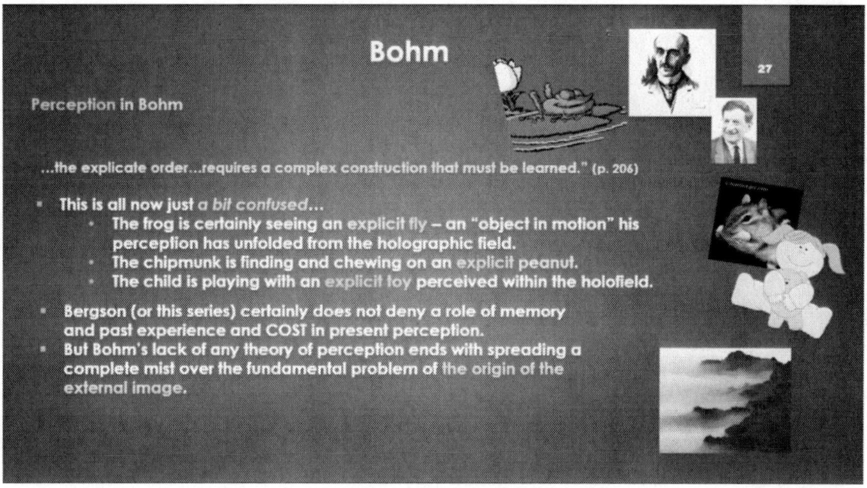

This is all now just a bit confused. As I noted, the frog is seeing an explicit fly – an "object in motion" -- his perception is unfolded from the holographic field; he doesn't need to spend two years learning causality, object, space and time to see that fly. The chipmunk -- same thing – he is finding and chewing on an explicit peanut. The child is playing with an explicit toy perceived within the holofield. You don't need two years of cognitive development to do that. You need it for other things, yes. Now Bergson certainly does not deny the role of memory and past experience and COST (causality, object, space, and time) in present perception. But Bohm's lack of any theory of perception ends with spreading a complete mist over the fundamental problem of the origin of the external image. He's just taking us into realms that are irrelevant to the initial problem, which is basically the origin of the image of the external world -- unfolding that from the holofield with all its qualia -- because it is all part of the hard problem, obviously – because there is a qualitative dynamically changing holofield part of which is being specified as a coffee cup or a fly buzzing by.

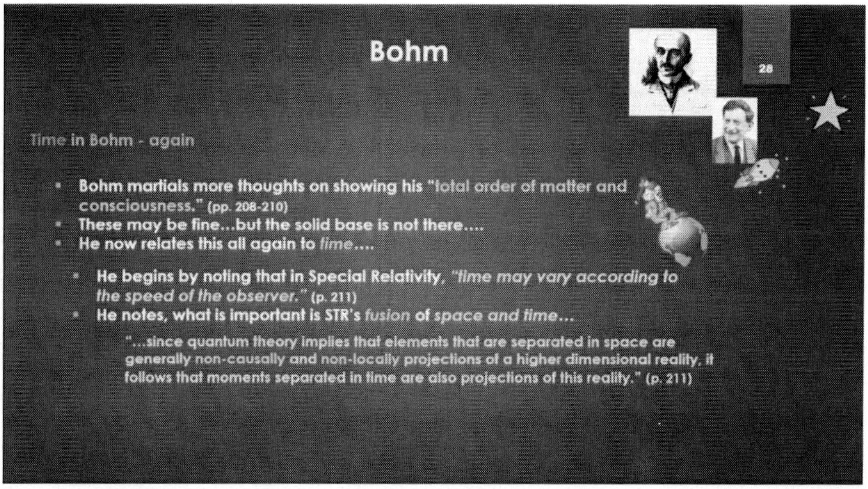

So, a little bit more on time in Bohm: he marshals more thoughts on showing his "**total order of matter and consciousness** (pp. 208-210)." These may be fine, but the solid base is just not there. He now relates this all again to time. He begins by noting that in Special Relativity, "**time may vary according to the speed of the observer** (p. 211)." He notes what is important is Special Relativity's **fusion** of **space and time**. And he says,

> . . .since quantum theory implies that elements are separated in space are elements that are generally **non-causally** and **non-locally** projections of a higher dimensional reality, it follows that moments separated in time are also projections of this reality. (p. 211)

So, "moments separated in time are also projections of this reality." So here we're going to get to this attempt to account for quantum jumps, quantum moments jumping to other quantum moments shall we say. And he never breaks from these mythologies of Special Relativity -- that time may vary according to the speed of the observer.

Pryzdia:
He never breaks from Special Relativity or the notion of quantum jumps.

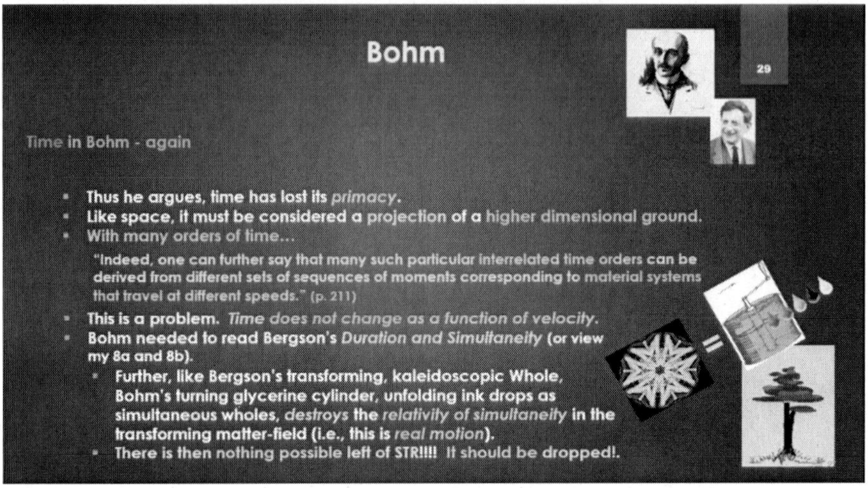

Steve:
Nope. Thus, he argues that time has lost its **primacy**. Like space, it must be considered a **projection** of a **higher dimensional ground. With many orders of time. . .**

> Indeed, one can further say that many such particular **interrelated time orders** can be derived from different sets of sequences of moments corresponding to material systems that travel **at different speeds**. (p. 211)

Well, this is a problem. **Time does not change as a function of velocity**. He needed to read Bergson's *Duration and Simultaneity* (or view my 8a and 8b). I know that to our audience, this statement is probably a jaw dropping flabbergasted statement [laughter] – that time does not change as a function of velocity. And it's a subject that you and I wanted to go into, but I don't know if we're going to be able . . .

Pryzdia:
Well, I just want to say your treatment of the Special Theory Relativity . . . for instance, here you refer to your #8a and #8b -- you're referring to your YouTube lectures . . . you do a

fabulous job critiquing the mythology around relativity. So, yeah . . .

Steve:
Yeah. And Bohm just. . . he needed to get through that. But he didn't. So further, like Bergson's transforming kaleidoscopic Whole . . . I think we talked about that a bit . . . Bergson said, yeah, ala relativity, one can ask, "Well, how do I know what object is in motion and what object is at rest?" But in reality, that's sort of an invalid question because you're identifying separate parts in the whole as though they had a reality when in fact it's the whole that's changing – as though it was a kaleidoscope (Bergson's metaphor) -- the whole is changing and now the motions of objects become changes or transferences of state -- like waves in an ocean, within the whole. And from that perspective, there's real motion, as I noted before: roses blooming, mountains arising, couch potatoes getting horribly fat, et cetera. There's real motion in the universe, and that real motion implies real simultaneity. And actually, the kaleidoscope changing is actually a simultaneous change across the entire universe. It's interesting because in Bohm's model of the turning glycerin symbol where the ink drops come out one by one by one as simultaneous wholes, he's really destroying the relativity of simultaneity in the transforming matter-field. That is, his glycerin symbol is a whole turning -- and outcoming these little particles. He admits that his cylinders with the glycerin embedded between two surfaces of the cylinder is . . . he doesn't like its mechanical implications, and he kind of wants to drop it -- because of course, it starts to go against other things he thinks are relevant in physics. But as I'm looking at my last sentence there -- if you're destroying the relativity of simultaneity, there's nothing possible left of Special Relativity; it should be dropped, but he can't go there.

Pryzdia:
But I just want to say something about the kaleidoscope there. I think he would like that as a metaphor.

Steve:
Yeah.

Pryzdia:
But I think, if you bring Bohm into this . . . it's almost as if that kaleidoscope is a **pulsating kaleidoscope** for Bohm.

Steve:
Exactly. Yeah. That's what I was going to say. But he would say that kaleidoscope is a series of quantum jumps -- jump after jump after jump after jump. And Bergson would say: no way.

Pryzdia:
That's the difference right there.

Steve:
It's interesting though, that Bergson never actually addressed the quantum jump problem, but I don't know that . . . and that was around before he left the planet, I think . . . 1927 or so.

Pryzdia:
He died in 1941.

Steve:
He just never addressed it. He might not have known what to do with it. You know?

Pryzdia:
Too bad too, by the way.

Steve:
Yeah, yeah. Would've helped me.

Pryzdia:
Me too.

Steve:
Some more real motion there (referring to the slide above) – an organically growing tree that can't be relativized. You can't actually relativize such a motion without destroying the organic growth -- the tree would fall apart. So anyhow . . .

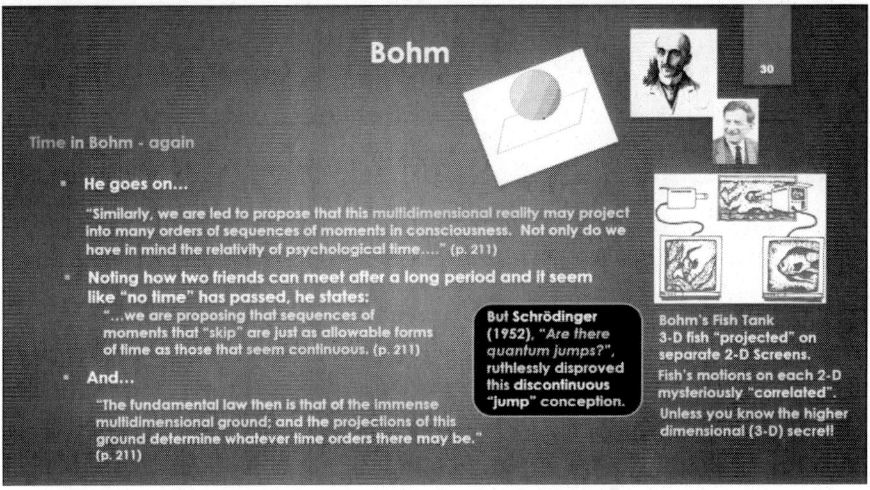

He goes on . . .

> Similarly, we are led to propose that this multidimensional reality may project into many orders of sequences of moments in consciousness. [Again, we're back to the "moments" in consciousness -- sequences of moments.] Not only do we have in mind the relativity of psychological time . . . (p. 211)

Remember (see above) he's got his fish tank model where different views of the fish are being projected depending on which camera you're looking at (e.g., facial view of the fish or the side view of the fish) -- but in reality, it's one fish. That is one of his models. He said that the fish motions of each 2-D fish are mysteriously "correlated." We've got this higher dimensional reality, maybe the fish in the tank, that's being projected down into a low dimensional reality, et cetera.

Or (referring to a graphic on the slide above) a little model up there . . . what it looks like if you're a flatlander living on that plane -- you've got an expanding circle in that plane; but in reality, it's a three-dimensional ball going through the plane, if you were able to look at it from a higher dimension. So, he's sort of thinking along these lines -- noting how two friends can meet after a long period and it seems like no time has passed, he states: " . . .we are proposing that sequences of moments that '**skip**' are just as allowable forms of time as those that **seem continuous** (p.211)." I would say that he's really getting into the weeds here -- bad weeds. Because he's just trying to deal with quantum jumps. Bohm goes on: "And the fundamental law then is that of the **immense multidimensional ground** and the **projections of this ground** determine whatever time orders there may be (p. 211)." So, he's visualizing this ground, which by the way, has to be continuous. I mean, at some point you're going to have a continuous generator of events, but your continuous generator of events can generate things as the movie slide or . . . movie screen . . . snapshot after snapshot, screen after screen. And obviously . . . by the way, it could present a picture of you when you were one, and then jump to a picture of you when you're ten. And that's allowable too, because this higher dimensional ground can project anything it feels like at any time. So . . .

Pryzdia:
Are we saying here that the multi-dimensional ground is the holomovement? And for him, time is in the holomovement. That's the issue with Bohm, right?

Steve:
Right. Well, yeah. But then again, his multidimensional ground . . . Does that come and go in discrete jumps too? What's creating the discrete jumps? You see, now you're back to Descartes -- looking at a discrete state model of time where each instant of time would correspond to say . . . visualizing a cube of the universe – the universe as a big cube, an instantaneous cube . . . You have cube one, then you have cube two, then you have cube three. The problem is they're frozen cubes. You can't get from cube one to cube two unless God creates cube two. Cube two disappears. But now you have to have God create cube three. I mean . . . Descartes literally said, well this is where you would need God to create the universe over and over and over again. Well, who's creating the multi-dimensional ground if it's a discreet state multi-dimensional ground? I mean . . .. it all goes back to these absurdities, unless, as Bergson said, motion is primary -- end of story -- you're always led back to movement being the fundamental thing.

Pryzdia:
But that's confusing to me. Bohm is also saying that motion is primary because for him, the holomovement is primary.

Steve:
Yeah. But you see . . . But yeah, I guess the holomovement would be primary, but . . .

Pryzdia:
And it includes time – time unfolds from the holomovement. So, it's upside down with Bergson. **For Bergson time is primary, and for Bohm order is primary.** See, that's the difference, I think. **Bergson is grounded in time and Bohm is grounded in order.** And that is a fundamental difference.

Steve:
I think so. I think we're getting to the . . ., oh, but my little comment here . . . (referring to the slide above)]: But Schrodinger in his paper in 1952, "Are there quantum jumps?" ruthlessly disproved this discontinuous "jump" conception.

Pryzdia:
Yeah. Nobody knows that.

Steve:
I just think he ripped the ground out from under Bohm.

Pryzdia:
Nobody talks about this though. Nobody talks about that 52 paper.

Steve:
No, no. Because the, this mythology of the quantum jumps and all these other quantum thingies is so mystically mysterious and fun that . . . Not that Bergson isn't plenty mystical, mysterious and fun -- but just a different form. Anyhow . . . this information is shared in my #60 presentation on YouTube ("Bergson's Holographic Theory").

Pryzdia:
Right.

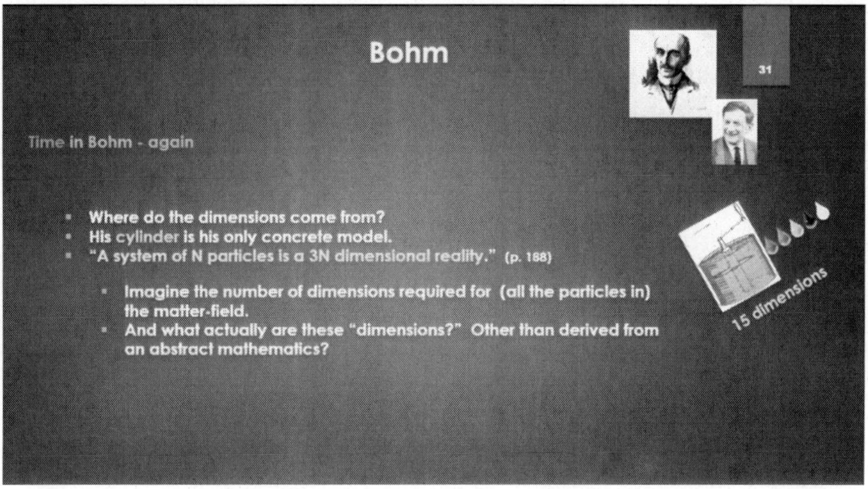

Steve:
Again, where do the dimensions come from? His cylinder is his only concrete model. Bohm states, "A system of N particles is a 3N dimensional reality (p.18)."

If I have four particles -- three times four -- I have twelve dimensions. I have five particles there coming out of the cylinder, but I need fifteen dimensions to do that. That's quite an interesting little cylinder. Okay. Imagine the number of dimensions required (for all the particles in) the matter-field. And what actually are these "dimensions" other than simply derived from an abstract mathematics? This is where his model sort of crashes into the concrete wall of concreteness, shall we say. And that is all I had to say about Bohm at this point.

Pryzdia:
Okay.

Steve:
Okay. I think we said enough though.

Pryzdia:
Well, yes, there's a lot there. And I want to say that I think

any scholar or any Bohm fan needs to watch your #60 presentation -- you just mentioned the one on Schrodinger and the quantum jumps. That's so crucial for people to understand. And then what you say about Special Relativity in #8a and #8b -- that needs to be . . . because, as you say, Bohm definitely "drank the Kool-Aid" with regard to the Special Theory, right? And he never let go of quantum jumps. And . . . he gave us good food for thought.

Steve:
Right. I would recommend the one I did on Whitehead too -- #53 -- because there you're looking at the notion of "epochs" or these "moments," et cetera, and how that just starts to fall apart -- and Whitehead's solution to the relativity problem. And I say solution in quotes because it's instructive that, again, it gets destroyed by Bergson's analysis. And #53 is even more succinct than #8a or #8b.

Pryzdia:
All right. So, you want to bring this to a close for now?

Steve:
Yeah, we can do that. All right, let's do that.

Pryzdia:
Signing off then.

Talk 3: On Bergson and Bohm, Part 1

[NOTE: The opening dialogue below representing the first twenty minutes of "A Bergson-Bohm Comparison (Talk 3: 'On Bergson and Bohm, Part 1')" does not appear in the edited Zoom recording; however, it contains valuable information that is alluded to later in this presentation and in the presentation that follows (i.e., Part 2).]

Pryzdia:
Hi Steve. I wanted to ask you a few questions before we actually begin. By the way, are you ready to do a presentation on "Bohm 2"? Can you do that this morning or do you want to wait?

Steve:
I think so.

Pryzdia:
All right. Well, before we do, let me ask you . . .there are two ways we can go about this. I still have two or three questions for you that relate to this presentation you are about to give. I didn't know if you want me to share them with you now, or if you would prefer me to wait to make a comment during your presentation. I kind of want to share questions beforehand so you kind of know what's on my mind. Can we do that real quick?

Steve:
Sure.

Pryzdia:
All right.

Steve:
Sounds good.

Pryzdia:
All right. So, what I've done since we last talked is . . . I think I wrote you an email a couple days ago . . . I went back and reviewed some of my old notes on some of this information that we're about to go over. For instance, I went over Shelly Joye's work, who I've been reading for years, and I went over Pribram. Additionally, I went over your YouTube video on Pribram (your #25) – which I have always felt is spot on. But here is what I wanted to read . . . I'll read a little from my notes and you can kind of just sit with this and meditate a bit on what I share in order to get an idea of where I'm coming from here. And I'm doing this, by the way, because I've been meditating for a while on all of this information and I'm feeling like I need to get comfortable with where I am with it all. That is, I need to make some conclusions with regard to your work and Bergson and Bohm. So, this is what I have here.

It seems to me that people writing about Bergson, Bohm, Whitehead . . . they are trying to integrate all of these folks . . . and it seems to me that they say something like the following. For most of these people, there are generally **two orders**. You can call it the space-time order, or the space-time domain. Call it what you want. Sometimes it's called the macro-level or the classical realm, or in Bohm's case, it is the explicate order. Whitehead talks about actual occasions or actual entities. And then you have the "beyond" space-time order or beyond the spacetime domain -- which sometimes is called the micro-level or the sub-quantum realm. Or in Bohm's case, it is the implicate order. Or in Pribram's case it is the spectral domain. Whitehead speaks of eternal objects. OK. So, for most of these people, there are two orders. And it could be said that for most of these people information somehow passes back and forth between the two orders or realms or domains (for Pribram, for example, it is by way of a Fourier transform).

Now . . . **true creation or true novelty**, for these people, **unfolds from beyond this space-time realm, order, or domain.** For example, Bohm speaks of an implicate order as a **generative order** within a **holomovement**. And then he speaks of reality (and for Bohm, reality is always defined as "what we think about"). He says reality consists of an infinity of levels that unfold as part of a holomovement. There's no absolute ground. There's active information and inactive information, grounded in **meaning** -- which by the way, it seems that most people don't meditate on that – for Bohm, everything is grounded in meaning – which fundamentally, in the end, is grounded in "**mystery**" – which, of course, makes him sound quite mystical in the end.

Now . . . for Bergson, there is **extension** and **tension** in or of **one order**, which he calls "duration." For Bergson, **time itself is creative. <u>It appears</u> that the holographic theory of Bergson does not require two orders**. And when I first discovered your work back in 2017 or 2018, that's the first thing that struck me. I asked myself: "How do you have a holographic theory without two orders?" And how do you have novelty just within time itself? So, I just wanted to read that to you. And here's another point. I came across another one of these Bergson anthologies, and I've just stopped sharing these anthologies with you because I have assumed that you have seen too many of them already [laughter], and don't really . . . this one's called *The Challenge of Bergsonism* (2003) by Leonard Lawler. I don't know if you're familiar with Leonard Lawler. He's written a little bit on Bergson.

Steve:
Yes. I'm familiar with him.

Pryzdia:
I really like this book -- *The Challenge of Bergsonism.* Right away, in the opening sentences of the Preface, he states that he is going to summarize Bergson's definition of duration. And this is what he says. And I do like this. He says, "duration equals memory plus the absolutely new."

Steve:
Absolutely what?

Pryzdia:
The absolutely new. Basically, duration is memory plus novelty. All right? Now based on my reading and understanding of Bergson, it's really hard to disagree with this definition as far as I can tell, but obviously memory and the absolutely new are not the same thing. They can't be. Thus, memory and novelty are of two orders. Correct or incorrect? Can someone please explain to me how you can say that duration is memory plus the absolutely new?

Now, two questions -- I need to know, and you have talked about this in your videos, this whole "extension/tension" thing that one finds in Bergson's work -- so here's a question: extension/tension; is this not properly a pulsation? Yes, or no? Now, it doesn't need to be. For example, I'm always picturing a rubber band where something is tensed up and relaxed, you know, and that doesn't need . . . there's no unfolding there. You tense, you relax. You know? So, here's a point. You're familiar with Gunther, of course, the editor of *Bergson and the Evolution of Physics* (1969). And you read Capek's book *Bergson and Modern Physics* (1971) last week. Now, here's what Gunther says – and he says this in *The New Bergson* (1999), that book that came out 20 years ago or so, another one of those Bergson anthologies -- Gunther says that for Bergson, matter is conceived as "pulses of energy" -- matter is "temporalized energy." And I'm quoting him here: for Bergson, matter, like mind, is comprised of "rhythms of duration." My question for you Steve: what's your opinion of those statements there where he is talking about pulses of energy . . . all that . . . I just can't get around extension-tension versus the pulsation thing. And I need to know before we begin . . . where are you on that? Because it seems like you're not . . . you don't go with the whole two orders notion.

Steve:
Well [laughter], the first thing I am going to say is that you are more of a pioneer in this whole subject than I am. [laughter]

Pryzdia:
Oh, you don't . . .

Steve:
Because I have not focused on this level of questioning. Okay. I'm down trying to explain . . .

Pryzdia:
I know . . .

Steve:
. . . the image of the external world and how memory works and how perception might work, and how we can have a theory of cognition and memory. Okay?

Pryzdia:
Yeah.

Steve:
And, and so I'm going to plead that . . . first of all, before I say anything and make a . . . of myself [laughter] that you're more into the subject than I am. It's an interesting subject. I just

haven't thought that much about it. And as you can see from the progression of my YouTube presentations, I didn't really even bother with *Creative Evolution* much until . . . like #47 or #46. And that's only because this gal . . . Ann Malaise wanted me to write a chapter for her book on biology and evolution. And I said, sure, I can do that . . . [laughter] uh . . . now how am I going to do that? You know? Well, I looked at it a bit because of that "Meditation on the Mousetrap" article, you know . . . and . . . and so this . . . I mean you're talking to a relatively neophyte person here on this particular subject that you're on. So, my . . . okay, so I'll say a few things, but you're going to have to take it with the grain of salt it deserves here from a guy who hasn't thought about it that much.

The Lawler . . . there's two things. The Lawler thing . . . Lawlor's characterization of duration being memory plus creation, right?

Pryzdia:
Memory plus the absolutely new. Yeah.

Steve:
Yeah. But it's . . . those aren't two things. They're not two demarcated things in Bergson's thought in reality.

Pryzdia:
No, they're not.

Steve:
He's mischaracterizing this, you know? That leads you astray, I think, because . . . duration itself is creativity. [laughter]

Pryzdia:
Yes, that is what Bergson is saying, right?

Steve:
Yes. And why does he say that? Only because the whole past is reflected in the current moment. Okay? Every instant is the

reflection of all preceding instants -- remember the melody metaphor where every current note is the reflection of all preceding notes. And in Bergson's mind, that's the only reason why the next instant can be creative. Okay? Because it is summing up all of this past, and it can . . . I mean you have to think about this for a while. But remember Bergson's statement . . . "time is this hesitation or it is nothing." That is, before each instant, we are kind of figuring out the next instant based on all that we are. But that duration, that flow is inherently memory -- all simultaneously. It is memory that allows for creation. That is, it is simultaneously creation. So, sticking the two together with a plus sign -- duration plus memory -- is already . . . it sort of sends you . . . unless you understand what that means, it could easily send you off in the wrong direction. So, I tend to go with your assessment that there aren't these two orders. There's just this . . . there's one big order [laughter]. And then I plead my neophyte status in really thinking about this thing at that point. Because it is a cosmic question that is important. But I've always, you know, like I've said, I've always . . . I've always . . .

Pryzdia:
Stayed away . . .

Steve:
Taken up the . . . I thought the cause is down more at the "hard problem" level of Chalmers and his misdirection -- straightening that out.

Pryzdia:
I think what's happened to me is that I have been infected by Jiddu Krishnamurti and his influence on Bohm's thinking. Because they really went into "creation." And you know, for both these guys, Krishnamurti and Bohm, there is something

called a "moving stillness." Okay? So, as you know, Bohm like . . .

Steve:
A moving what?

Pryzdia:
A moving stillness – or a stillness which is in movement.

Steve:
Oh, moving stillness. Okay.

Pryzdia:
Which has to do with . . .

Steve:
Wait, you almost sent me into a Zen satori there.

Pryzdia:
[laughter] Well, there you go. See, for Krishnamurti . . . one of the most famous Bohm/Krishnamurti dialogues is called *The Ending of Time* (1985, 2014). And the point they make there is, **there is this possibility of a stillness which is in movement or a "moving stillness" -- from which intelligence and creation comes**. So, I really have a hard time with saying memory and creation go together -- because they're always making distinctions like . . . okay . . .. there's "**invention**," which is **in time**. Invention means you take something which already exists and move it around and do something new with it. But that's not "**creation**." Creation must come out of, well, to use a term that Bergson despises – **nothingness** – or, in their case, stillness – a nothingness, a stillness -- but which is **in movement**! Because, as you know, for Bohm, everything is grounded in the holomovement. So, I have a real hard time, and this is just me because I've been so infected with this Krishnamurti/Bohm material . . . I have a hard time with memory being also

somehow creation. I can see "time is inventive," but "time is creative"? I still have a hell of a time with that -- and it looks like I'm not alone. See, the other point is this whole notion of pulses of energy or a physical duration -- which pulses in and out or as Bohm . . . even Bergson speaks of "unfolding" -- duration unfolds matter (which seems to indicate, as I have shared with you before, that Bergson's "duration" can be perceived to be a Bohmian "generative order"). So, I can't get around this pulsation thing either. Those are the two things that bother me. Time is creative -- and I have a hell of a time not going with . . . there has to be some kind of a **pulsating energy** here, a **pulsating matter** here -- which is, of course, a fundamental part of Bohm's physics -- this whole enfolding-unfolding thing.

So, to have a holographic theory, and I'm not saying . . . I mean Bergson is just "blowing my mind" because . . . There's a holographic theory which does not require two orders. There's no enfolding-unfolding required here. I mean, it's brilliant! But it's really difficult for a guy like me [laughter] to get my . . . and most likely anybody else who has studied Bohm . . . I'm glad we're doing this Bohm presentation today because anybody who looks at Bohm and then comes across a holographic theory without two orders is going to be somewhat shocked. It's like, wow, how do you do that? You know?

Steve:
That's an amazing observation. Because I would never have thought that, but I can sense your . . . you know . . . it's interesting that you say that. Yeah.

Pryzdia:
Yeah. What I love . . .

Steve:
It just didn't dawn on me that that would be the case. [laughter]

Pryzdia:
What I love about what we're about to do . . . I think what we need to do is get to your work. Because, as you know, I have listed a whole bunch of things that hopefully we can come across or talk about at some point. But what's great about what you're doing though is how you are bringing in J.J. Gibson's work, bringing in the scale of time issue, bringing in your critique of the reception of the Special Theory of Relativity, Susskind's' holographic theory and how much it differs from that of Bergson and Bohm, the quantum jumps, all of that – we have to bring that content out. How do you look at Bohm with regard to Bergson's holographic theory – how do you put the two next to each other? You have said to me before that when it comes to publishing, you try to stay away from some of the more mystical content. Not that you're not interested in mystical content personally, but you've said to me before that you prefer not to focus on that in your writings.

Steve:
Right. And the reason being is that Priority A is to bring science back to where it belongs, especially the science of consciousness, cognition, and perception. Physics of course has got to follow too. I have come to that point where I am now wondering if . . . I now reject the whole notion that everything has to be based on physics. Because in current physics, you can't base anything on – it is itself based in the wrong metaphysics. It is not a foundation that you can base anything on. You can't base a theory of consciousness and perception on current physics. It just isn't going to work. But I've always thought if you could move people's perception of

the brain back into the Bergsonian framework, then you've got a better foundation for the mystical. And I have always thought that that's the problem with people like . . . like a Lynn McTaggart or a Michael Talbot, etc. They're starting off at this other level, you know . . . throw in a few tofu blocks and claim that we've got this conception that affords the mystical; and they haven't done any of the underlying work necessary to make that work – so that . . . the whole scientific level just says, yeah, okay, that's nice but . . .

Pryzdia:
Even the whole notion that the holographic image is "projected" after it's "downloaded" -- all of that, right?

Steve:
Yes.

Pryzdia:
Alright. So . . . thanks for weighing in on that though, because it was weighing on my mind. I didn't know how much you've thought about what I just shared with you, so I really appreciate . . .

Steve:
Right.

Pryzdia:
I know where you're coming from.

Steve:
Yeah. Maybe when I'm an old man, which is probably about 177, I will eventually get to that question [laughter] – but by then I won't have any brain cells left to even think about it. [laughter]

Pryzdia:
Well, are you ready to do this Bohm presentation? Do you

want to do it? We could do this in pieces and of course we can edit whatever we like.

Steve:
Yeah. We can at least get some pieces . . . the first piece going.

Pryzdia:
Sure.

Steve:
Maybe we will get through the whole thing; it depends on what kind of a roll I get on here . . .

[Edited Zoom recoding below:]

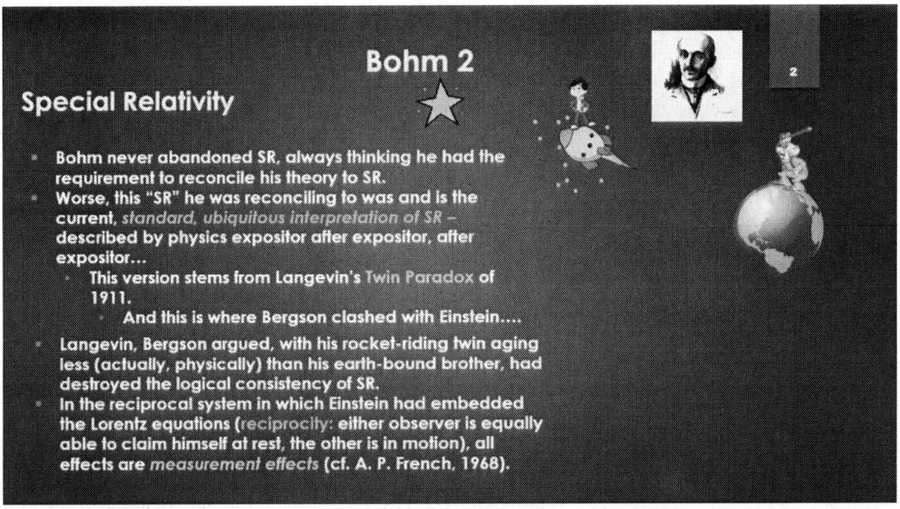

Pryzdia:
Okay. Hey Steve I'm glad we're back. I'm glad that you have agreed to move us further forward with this . . . I don't know what we're calling this . . . a Bergson-Bohm comparison or integration. We'll call it Bohm 2 -- following up on what we recorded a while back. So . . . I guess just take it away Steve; I know you've created a presentation here of some slides. Let's just move forward.

Steve:

Okay, this goes for a little bit here -- because we have a lot to look at in Bohm, some of which we looked at already. So going back to the Special Relativity subject first of all, the thing is Bohm never abandoned Special Relativity. And by that I mean the Special Relativity that is the current view of Special Relativity -- that is, that Special Relativity explains the twin paradox. It explains why the twin on the rocket would age less than his brother who is left on the earth, the old guy physically aged with a cane, et cetera. That's the ubiquitous version of Special Relativity right now. And that's what he was trying to reconcile his theory to. So, as I said, not only was he trying to reconcile to Special Relativity, he was trying to reconcile to the standard ubiquitous interpretation of Special Relativity -- that it explains things like the aging twin, the muon that's aging less rapidly as its velocity increases, the slowed clocks on the jets that are going around the planet like in the Hafele and Keating experiment, et cetera. So, as I said, this comes from Langevin's Twin Paradox of 1911 -- this is where Bergson clashed with Einstein.

In the argument, Bergson argued that Langevin's rocket riding twin was aging less (actually, very physically) than his earth-bound brother – this thinking had destroyed the logical consistency of Special Relativity -- and that Einstein had created a reciprocal system. Remember, either person could declare himself at rest and the other guy in motion. That is what reciprocity is defined as. Einstein had embedded the Lorentz transformations or equations in this reciprocity or reciprocal system. And by that then all effects are what are called "**measurement effects**." A.P. French in his book *Special Relativity* (1968) called them such. What's a measurement effect?

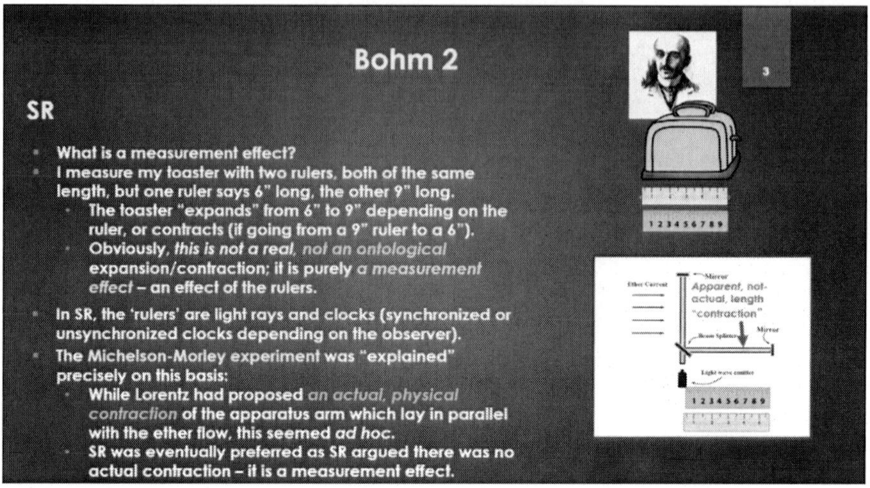

Well (see above), I measure my toaster with two rulers and both are exactly the same length, but they are mis-marked shall we say: one is six inches long, one is nine inches long. So, depending on which ruler I use, the toaster expands from six inches then to nine inches or contracts from nine inches to six inches.

This is obviously **not an ontological** effect, it's just purely a **measurement effect**, an effect of the rulers; the toaster isn't expanding or contracting. In Special Relativity, the "rulers" are light rays and clocks. It's the same thing. And the rulers change depending on whether the clocks are synchronized or unsynchronized -- depending on what the observer thinks. A stationary observer thinks his clocks are synchronized; an observer flying by him who happens to think that he is at rest and the first observer is in motion says, "no, your clocks aren't synchronized." And so that's where the equivalent of the expanding or contracting rulers comes in in Special Relativity. So it was on this basis that the Michelson-Morley experiment was explained.

The experiment had two arms and one was parallel to the ether current and the other one was orthogonal to the ether current as the picture shows there (see above). Lorentz proposed that there was **an actual physical contraction** of the apparatus arm that lay parallel with the ether flow. And that allowed the two light rays that were sent to each end and back to come back at precisely at the same time. By the Pythagorean theorem, one could compute that the light ray that was traveling in parallel with the ether should take a little bit longer than the one going orthogonal to the ether current. The problem was that physics didn't like that. Even though Lorentz had proposed electrodynamical arguments for this contraction, it seemed, according to the story, ad hoc to physics. And Special Relativity was preferred because there was no actual contraction of that apparatus arm. It was just a measurement effect. So, everything depends, in explaining the Michelson-Morley experiment, on the interpretation of Special Relativity -- that it is doing merely measurement effects like our toaster situation. There is not an actual length contraction -- that's the idea --- it is a measurement effect.

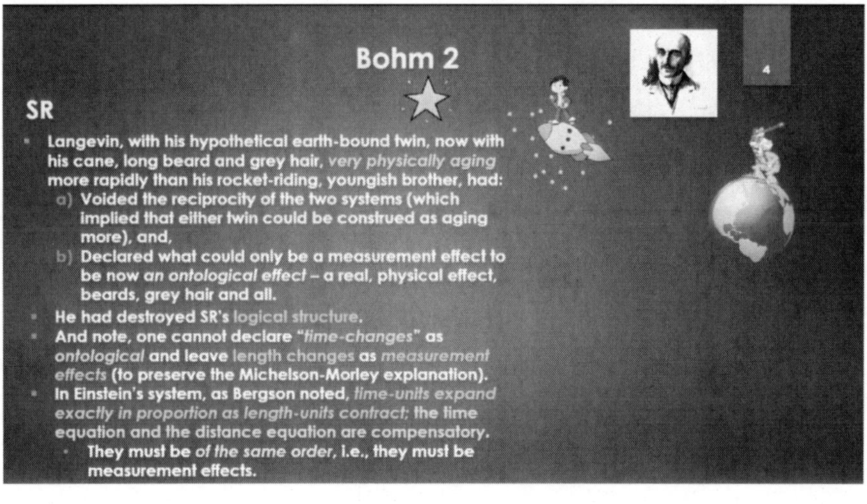

So (see above) when Langevin came along in 1911 with this hypothetical earthbound twin with his cane, long beard, grey hair -- **very physically aging** -- more rapidly than his rocket-riding youngish brother had -- Bergson argued that Langevin had voided the reciprocity of the two systems (which implied that either twin could be construed as aging more) -- you can just simply switch those systems and the earth could be moving away from the rocket and then back again – and now the guy on the earth is the younger twin. So basically, he declared that what could only be a measurement effect to be now **an ontological effect** -- a very real physical effect, beards, grey hair, and all. He had destroyed Special Relativity's **logical structure**.

This is Bergson's argument. But note: one can't declare the "**time changes**," the aging, as **ontological** and leave the **length changes** as **measurement effects** (in order to preserve the Michaelson-Morley explanation). And why is that? It is because in Einstein's system, as Bergson noted, **time-units expand exactly in proportion of as length-units contract; the time and distance equations are compensatory**. So, they have to be **of the same order**; they must be measurement effects or your whole explanation for Michelson-Morley dies and you're back to the existence of the ether, which is probably where we should be.

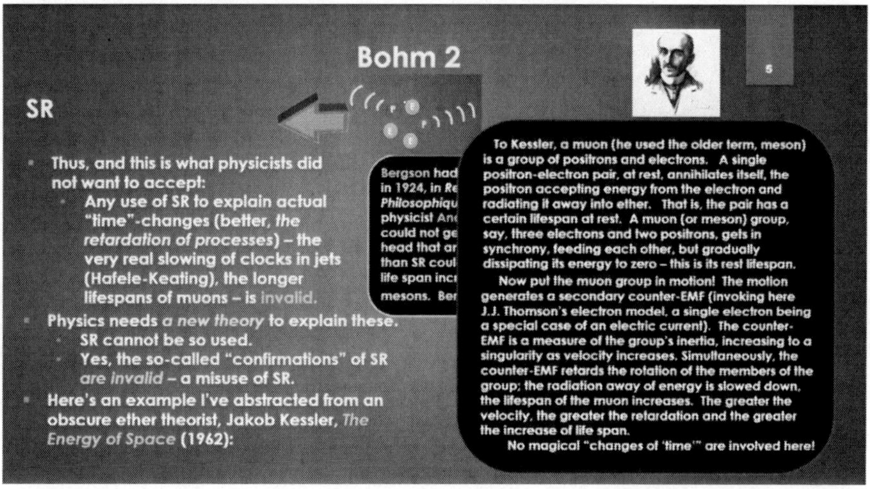

Thus, and this is what physicists did not want to accept: Any use of Special Relativity to explain actual "time" changes -- and I put time in quotes (see above) because you're not really explaining time changes at all -- that's complete mythology -- is better, stated "**the retardation of processes**" -- and that's not equivalent to the changes of time -- the very real slowing of clocks in jets (like Hafele-Keating), the longer life-spans of muons – is invalid. You can't use Special Relativity to explain real ontological physical effects, only measurement effects, only the rulers for the toaster. Bergson had this exact argument in 1924, in *Revue Philosophique*, with physicist Andre Metz. And Metz just could not get through his head that anything other than Special Relativity could explain the lifespan increases of mesons (that's the old terminology for muons). Bergson gave up. Physics needs a new theory to explain these said Bergson. Special Relativity can't be used. All the so-called "confirmations" of Special Relativity -- the "stunning confirmations" of Special Relativity that Jimena Canales refers to in her book *The Physicist and the Philosopher* (2013) – **are invalid** – they are a misuse of Special Relativity.

Here's an example (see above): I have abstracted from an obscure theorist, Jakob Kessler, *The Energy of Space* (1962) – a self-published book. This is to help someone like Metz or these physicists who can't grasp how there could be a different explanation that doesn't involve Special Relativity for those slowing muons. To Kessler, a muon (he used the older term, meson) is a group of positrons and electrons. (So, I've got a little group of positrons and electrons up there.) A single positron-electron pair, at rest, annihilates itself, the positron accepting energy from the electron and radiating it away into the ether. That is, the pair, has a certain lifespan at rest. A muon (or meson) group, say, three electrons and two positrons, gets in synchrony, feeding each other, but gradually dissipating its energy to zero -- this is its rest lifespan.

But now put the whole muon group in motion! Now the motion generates a secondary counter electromotive force (and I'm evoking here J.J. Thompson's electron model, where he saw a single electron being a special case of an electric current). The counter-EMF is a measure of the group's inertia, increasing to a singularity as velocity increases. Simultaneously, the counter (or secondary) EMF retards the rotation of the members of the group; the radiation away of energy has slowed down, the lifespan of the muon increases. The greater the velocity, the greater the retardation and the greater the increase of lifespan. No magical "changes of 'time'" are involved here.

You don't need Special Relativity. In other words, this would be a theory based in the ether that explains why the muons have a longer lifespan with greater velocity -- or the clock slow down with greater velocity. Or maybe even the twin would slow down a bit and age less, more slowly. But this has nothing to do with Special Relativity which can't be so used.

I don't think I've got it said here, but this assumes, by the way, clarity on a fundamental confusion that reigns throughout this literature, because the expositors of Special Relativity will immediately jump to the General Theory and accelerations to explain all this. They will say [Steve is referring here to an illustration that he used to explain Whitehead's view], well because the twin was accelerated in a rocket and had to turn around at the star . . . and then come back and accelerate to do so, this explains it. Well . . . but then you've got "stunning confirmations" of something in the General Theory, but something that has nothing to do with the Special Relativity theory. You haven't confirmed jack as far as the Special Relativity theory. And this is an utter misdirected illusion -- that appealing to accelerations is somehow stunning confirmation of Special Relativity.

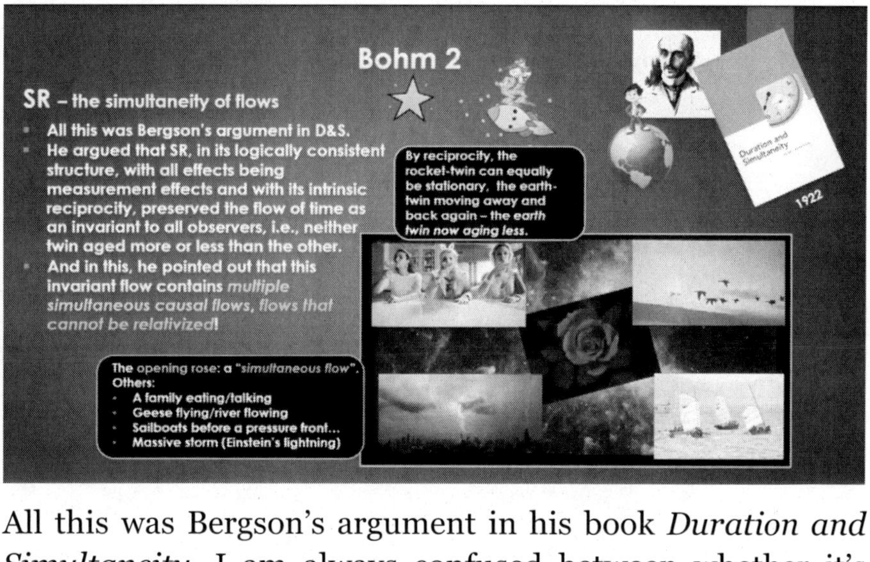

All this was Bergson's argument in his book *Duration and Simultaneity*. I am always confused between whether it's 1922 or 1923, but one of those two years [the book was published in 1922] -- he argued that Special Relativity and its logically consistent structure, with all effects being measurement effects and with an intrinsic reciprocity, preserved the flow of time as an invariant to all observers, and that neither twin aged more or less than the other. Again, by reciprocity -- the rocket-twin can be equally considered to be stationary, the earth twin moving away and back again. So now the earth twin is aging less, just reverse the situation. And in this argument in *Duration and Simultaneity*, he pointed out that this invariant flow, the invariant flow of time to both observers, contains **multiple simultaneous causal flows, flows that cannot be relativized**!

So, I'm picturing some here (see above), like the opening of that rose -- a simultaneous causal flow – an organic growth is involved. If you visualize all the points exploding outwards in space so to speak, that is a simultaneous flow -- or a family eating and talking -- or geese flying over the river, which is also flowing -- or sailboats before a pressure front -- or even the iconic lightning bolts of Einstein coming from what is, in effect, a simultaneous flow -- a massive storm.

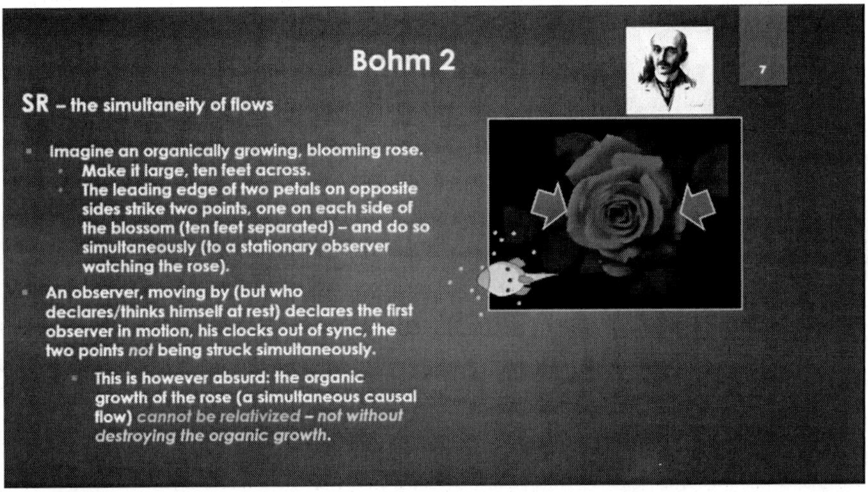

Imagine for example an organically growing, blooming rose (see above). Make it a large rose, about ten feet across. The leading edge of two petals on opposite sides strike two points, one on each side of the blossom (ten feet separated) -- and they do so simultaneously (to a stationary observer watching the rose). An observer, moving by (but who declares/thinks himself at rest) declares the first observer to be in motion, and his clocks out of sync (we're back to the ruler problem), and the two points are **not** being struck simultaneously. But this is absurd: the organic growth of the rose (a simultaneous causal flow) **cannot be relativized -- not without destroying the organic growth**.

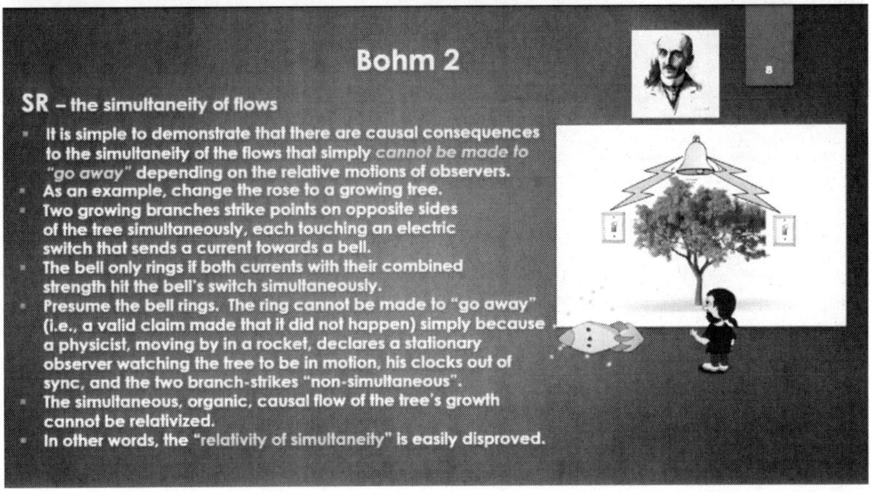

It's simple to demonstrate that there are causal consequences to the simultaneity of flows that simply **cannot be made to "go away"** – in other words, say they didn't happen depending on the relative motions of observers. As an example (see above), let's change the rose to a growing tree. Two growing branches strike two points on opposite sides of the tree simultaneously, each branch touching an electric switch that sends a current towards a bell. The bell only rings if both currents with their combined strength hit the bell's switch simultaneously. Presume the bell rings. The ring can't be made to "go away" (i.e., a valid claim made that it didn't happen) simply because a physicist/an observer, moving by in a rocket, declares a stationary observer watching the tree to be in motion, his clocks out of sync, and the two branch-strikes "non-simultaneous." The simultaneous, organic, causal flow of the tree's growth cannot be relativized. In other words, the "**relativity of simultaneity**" is easily disproved.

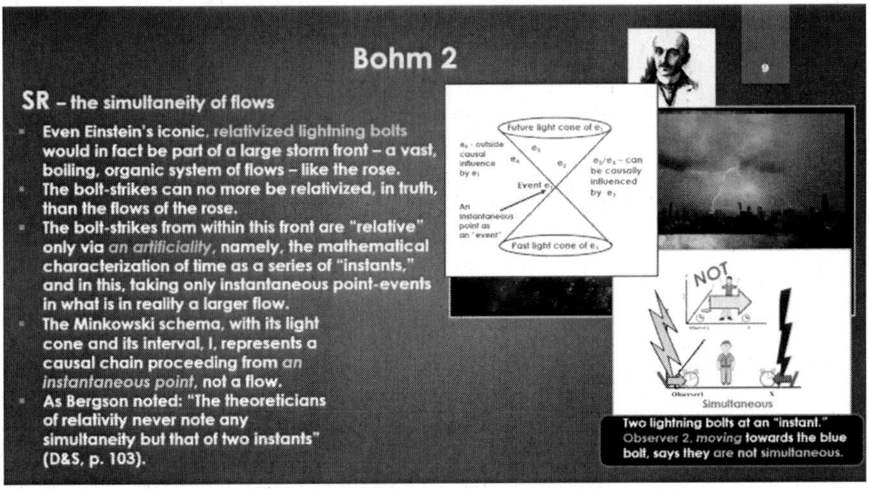

Even Einstein's iconic, **relativized lightning bolts** would in fact be part of a large storm front -- a vast boiling, organic system of flows -- like the rose. The bolt strikes can no more be relativized, in truth, than the flows of the rose. The bolt-strikes from within this front are "relative" only via **an artificiality**, namely, the mathematical characterization of time as a series of "instants," and in this, taking only instantaneous point-events in what is in reality a larger flow.

In my little standard relativity picture there (see above), I'm taking two lightning bolts at an "instant." Observer 2, moving towards the blue lightning bolt, says they **are not simultaneous** because the Green Observer (Observer 1) . . . his clocks are not simultaneous.

But the Minkowski schema (see above), with its light cone and its interval, "I," represents a causal chain proceeding from **an instantaneous point**, not a flow. Remember the light cones of the Minkowski schema. Everything from event e1 there in that cone, like events e2 and e3, can be causally influenced by e1 because they're reachable at the speed of light or less. E4 to the left there is too far outside that light cone. It can't be reached in time at the speed of light for e1 to influence it. But everything is proceeding from that point, not from a flow. And Bergson noted in *Duration and Simultaneity*: "The theoreticians of relativity never note any simultaneity but that of two "instants (p. 103)."

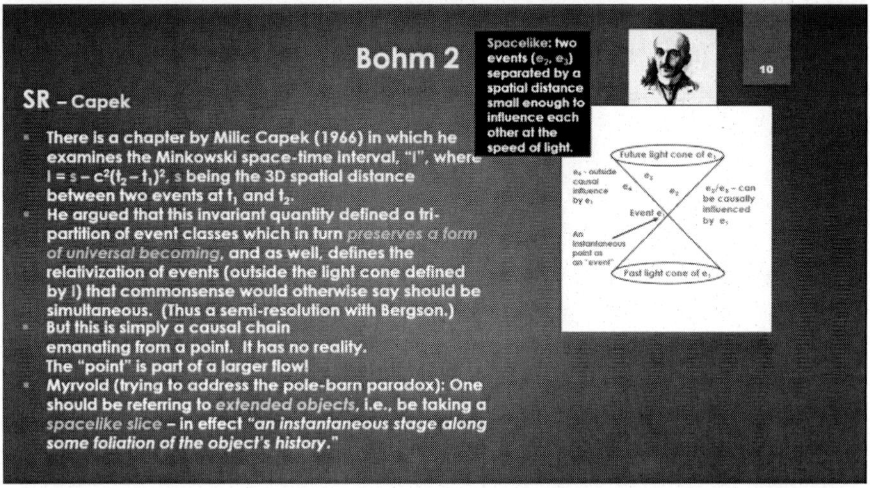

There's a chapter by Capek (1966) in which he examines the Minkowski space- time interval (see above), "I," where I = s - c2 (t2-t1)2, s being the 3D spatial distance between two events at t1 and t2. He argued that this invariant quantity defined a tri-partition of event classes which in turn **preserves a form of universal becoming**, and as well, defines the relativization of events (outside the light cone defined by I) that commonsense would otherwise say should be simultaneous. (Thus, a semi-resolution with Bergson.) As I noted before, the events inside the light cone (e3 and e2) can't be relativized because they're all part of a causal chain proceeding from that point. E4 exists outside the light cone; so, it can be relativized. We don't know whether it actually occurred before e3 or before e4 or simultaneous with et cetera. So those events outside the light cone are relativized. But according to Capek, there is a becoming for everything inside the light cone from that point; there is becoming, which we know Bergson insisted on -- that there is becoming and the universe is preserved by this form, this vision of relativity. Though you can get easily described situations where e4 would have to be influenced by e1 and also simultaneous, say with another event there inside the light cone. Anyhow, this seems to be Capek's attempt to reconcile Special Relativity to Bergson. But again, this is simply a causal chain emanating from a point. It has no reality. The "point" is actually part of a larger flow! In other words, if we were to take that rose and slap it right over the intersection there where e1 is, you'd have a much longer line of points initially. Take the starting configuration of the rose, you have a whole long series of points, an extended object that is going to unfold over time. And this is what Myrvold tried to say. He was looking at the pole-barn paradox. And he said: one should be referring to **extended objects**, i.e., be taking a **space-like slice** – in effect "an instantaneous stage along some foliation of the object's history."

"Spacelike" means: two events (like e2 and e3) are separated by a spatial distance small enough to influence each other at the speed of light. Or e1 and e2 are close enough spatially to influence each other. But now under Myrvold's thing, we're just including the outer petals of the rose as part of **a causal flow**.

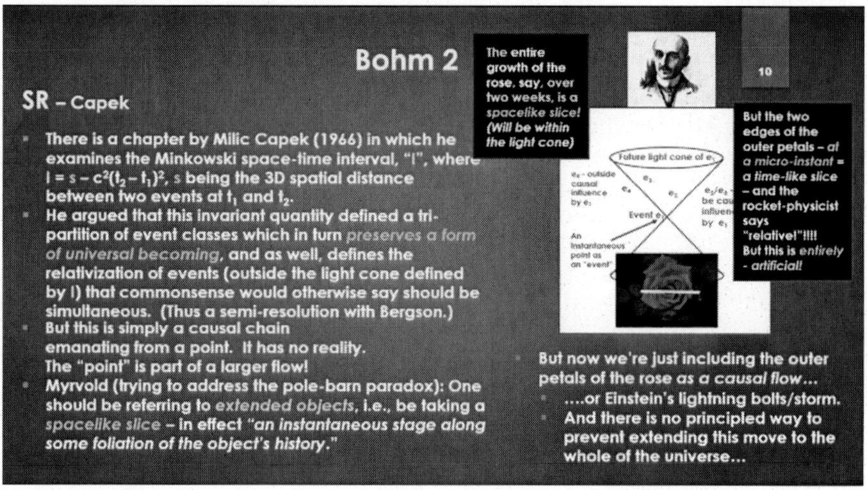

And you could take the growth of the rose (going back to its natural speed) over two weeks and it's all a nice space-like slice over two weeks. It's all in the light cone and it can't be relativized (see above). It can be only relativized if you take the two edges of those two outer petals of the rose **at a micro instant** -- now you've got **a time-like slice** – and the rocket physicist says, well "they're relative!" But this is **entirely artificial**! You have simply gone back to ignoring the whole flow, the causal flow of the rose. And you are basically saying that we're simply going to look at this micro instant of time where these two outer petals (one on each side) hit the light switch shall we say. And it is declared "oh, they're relative." But again, that's entirely artificial. It is just like Einstein's lightning bolts from the storm. There is simply no principle principled way to avoid extending this move of Myrvold's to the whole of the universe -- and saying, well the whole thing is a causal flow -- the whole transformation of the universal field is one of these simultaneous causal flows. And that's really what Bergson implied in *Matter of Memory* even before he wrote *Duration and Simultaneity*.

Pryzdia:
I just want to say, focusing on that micro instant, that is what Capek . . . in doing that, he makes his case.

Steve:
Right.

Pryzdia:
I'm suggesting that that is not what Bergson . . .

Steve:
Right. Exactly. So, let's sum up. Why did we say this? Basically, to un-shock the listener or reader from the idea that Bohm was completely misguided in trying to reconcile his theory constantly with relativity. And we're going to see this as we go on shortly, that he is always referring back to Special Relativity and the idea that times can be different -- Clock A can be going slower than Clock B depending on the velocity of the clock, et cetera. He's completely buying the BS of current Special Relativity and its current ubiquitous exposition by every expositor I've ever seen under the sun – they mindlessly repeat this over and over. They utterly fail to grasp what's going on, including Capek. Capek took this position against Bergson as well in his whole argument. He invoked the notion that Bergson didn't understand accelerations. Sorry, that's a complete myth. So that's the point -- we are going to be seeing Bohm always referring to Special Relativity, deferring to Special Relativity. This is such a fundamental misdirect that he's accepted and it makes it very difficult for me to go deeper into considering some of his theory.

Pryzdia:
As you may or may not know, he really did struggle with his attempt to reconcile relativity with quantum mechanics.

Steve:
And it's all because of this stupid mind problem.

Pryzdia:
Yes. I get it.

Steve:
So anyhow, I hope we made that point.

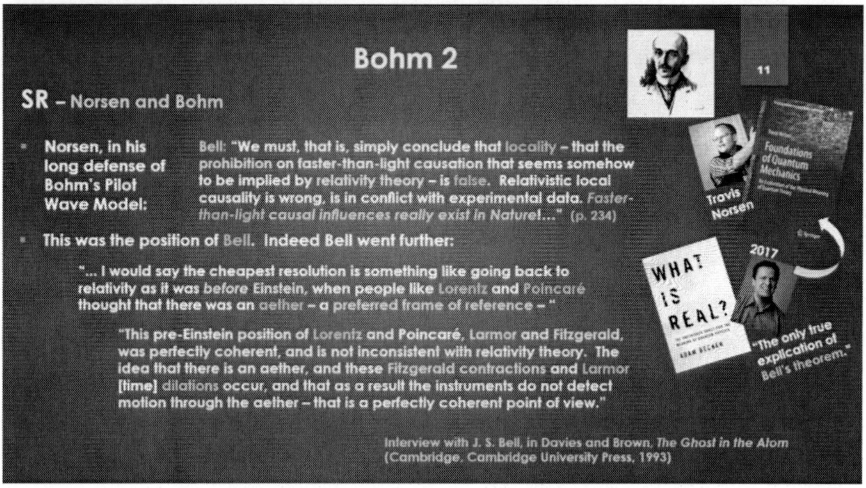

Now, Norsen on Bohm . . . Norsen remember wrote a book . . . I don't remember . . . but he wrote his book on quantum mechanics and it is a very interesting book . . .

Pryzdia:
Foundations of Quantum Mechanics (2017).

Steve:
Right. Norsen was a great defender of Bohm.

Steve:
And a great . . . well the whole book was an exposition of Bell's theorem, Bell himself being a great defender of Bohm. Bell loved Bohm's pilot wave model. Here Norsen is quoting Bell on Bohm's pilot wave model:

We must, that is, simply conclude that **locality** – that the **prohibition on faster-than-light causation** that seems somehow to be implied by **relativity theory** -- is **false**. Relativistic local causality is wrong, is in conflict with experimental data. **Faster-than-light causal influences really exist in nature**! (p.234)

Contrary to the whole light cone picture that we are just talking about. This was the position of Bell. Indeed, Bell went further:

> I would say the cheapest resolution is something like going back to relativity as it was **before Einstein**, when people like Lorentz and Poincarè thought there was an ether -- a **preferred frame of reference** –
>
> This pre-Einstein position of Lorentz and Poincarè, Larmor and Fitzgerald, [Remember Lorentz and Fitzgerald and the contraction hypothesis -- which physics rejected because it was supposedly ad hoc according to this story -- and I emphasize "story."] was perfectly coherent, and is not inconsistent with the relativity theory. [But you can't be talking Special Relativity because Special Relativity doesn't allow for any real ontological effects. And Lorentz, Larmor and Fitzgerald got real ontological effects.] The idea that there is an ether, and these Fitzgerald contractions and Larmor time dilations occur, and that as a result the instruments do not detect motion through the ether – that is a perfectly coherent point of view. (Interview with J.S. Bell, in Davies and Brown, *The Ghost in the Atom*, Cambridge, Cambridge University Press, 1993)

Note that Adam Becker in *What Is Real?* (2018) noted that Norsen's book *Foundations of Quantum Mechanics* contains the only true explication of Bell's theorem, which I thought was interesting.

Pryzdia:
That Becker book is quite good by the way.

Steve:
And it's why I poured over in Norsen.

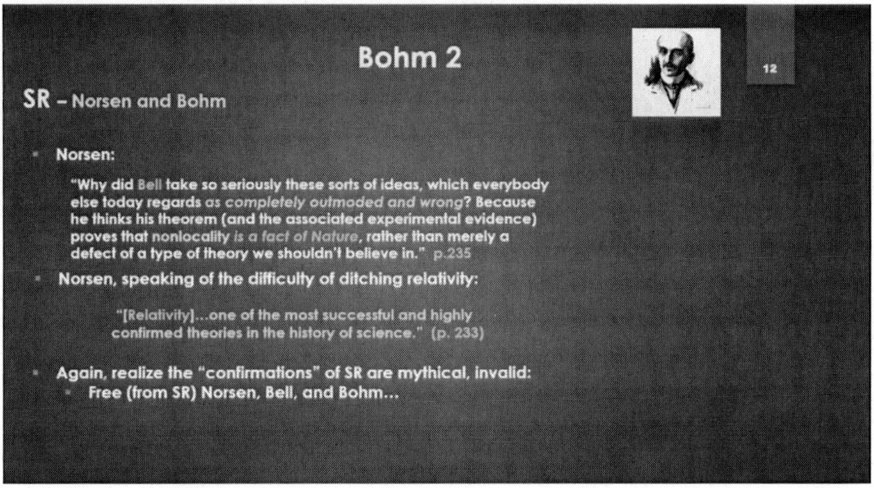

So, Norsen asks:

> Why did Bell take so seriously these sorts of ideas, which everybody else today regards as completely outmoded and wrong? Because he thinks his theorem (and the associated experimental evidence) proves that **non-locality is a fact of Nature**, rather than merely a defect of a type of theory we shouldn't believe. (p.235)

Norsen speaking of the difficulty of ditching relativity: "[Relativity] . . . one of the most successful and highly confirmed theories in the history of science (p.233)." Contemplate that sentence for a second. What have we just been talking about for the first ten slides? It's **not** confirmed. There is nothing confirming Special Relativity. It's complete mythology that it is so. So again, we're just frustratingly . . . We are always flustered by these guys just thinking this is an iron stake in the ground that can't be moved. It's ridiculous.

As I said, again, we need to realize that the "confirmations" are mythical, invalid. We need to free these guys (Norsen, Bell, Bohm) from this thing.

Okay, so that's bringing us to another . . .

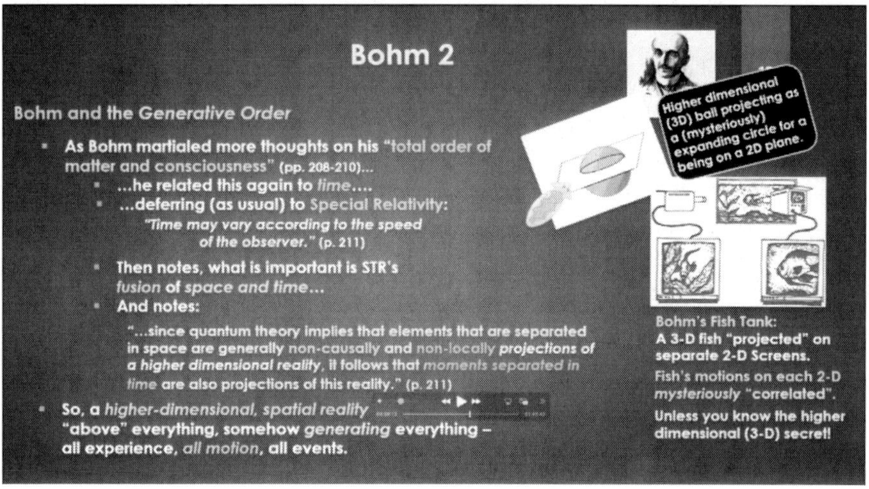

Pryzdia:
Yes, you're moving to a new topic here -- the Generative Order. I just wanted to say though, that the point you make in your work is . . . I like your phrase, "everybody drank the Kool-Aid" with regard to STR, including Bohm, and there's no doubt about it.

Steve:
Right.

Pryzdia:
It just shocked me. It's just shocking . . . the extent to . . .

Steve:
Even though Norsen and Bell are totally behind Bohm and this pilot wave model, they themselves are getting stuck with this mythology of relativity. So, we come to "Bohm and the

Generative Order." As he martialed more thoughts on his "**total order of matter and consciousness**," he related this again to **time** . . . again, deferring as usual to Special Relativity: "**Time may vary according to the speed of the observer** (p.211)." Then he notes that what is important is Special Theory of Relativity's **fusion of space and time**. Again, he keeps deferring to this complete freaky myth. And notes:

> . . .since quantum theory implies that elements that are separated in space are generally **non-causally** and **non-locally projections of a higher dimensional reality**, it follows that **moments separated in time** are also projections of this reality. (p. 211)

Now. . . But they're not . . . As you intuited, this is a non-temporal order. There's no real time in this order. The intuitive model is what we're looking at there (see above), where the flat two-dimensional plane being -- like a flatworm there watching an expanding circle on his plane, on his 2D plane of existence. And it seems like it's expanding. He has no idea why. And the reason it's expanding is because there's a three-dimensional sphere going through the two-dimensional plane. So, what appears like motion in two dimensions for the flat planaria is actually what is a solid 3D object, a sphere with no motion in and of itself, though it is moving through this plane, through the 2D plane. Which should already make you already kind of question the model because there this then this motion of the sphere going vertically up and down. Where does that come from?

Pryzdia:

Can I ask a question on this Steve?. . .

Steve:

Bohm's model is that it's just the sphere. But wait, the sphere is moving through the plane. How does that . . . it's not really coherent.

Pryzdia:
Well, does Bohm really need relativity to make . . . when he talks about the generative order, and he is . . . he's talking about quantum physics, he doesn't need Special Relativity to make the whole enfolding-unfolding/generative order work, right? I mean that's not important for that aspect of his work necessarily, is it? I mean . . .

Steve:
Yeah, it is. Because as is stated in that quote there "time may vary according to the speed of the observer."

Pryzdia:
I see what he said.

Steve:
It is folded into this whole notion that there's different orders of time, different speeds of time in his static generative order – there are different orders and speeds of time in that sphere that's now moving through the plane.

Pryzdia:
OK.

Steve:
So, any particular state of the expansion of that circle, the black circle on the two-dimensional plane would be moments separated in time, but they're just projections of that sphere

going through the plane according to Bohm (see above).[1] And yet somehow . . . the projection itself is a time, isn't it?. . . is motion . . . is becoming. You see . . . my struggle with Bohm and his models is that they look like they have coherence but they don't.

So, let's go on. A **higher-dimensional, spatial reality** seems to sit "above" everything, somehow **generating** everything -- all experience, **all motion**, all events. But yet somehow beneath the scenes it itself has to be moving through some other plane, like our 2D plane. I mean coherence starts to dissolve in my opinion.

Pryzdia:
I have to think about this some more. Go ahead.

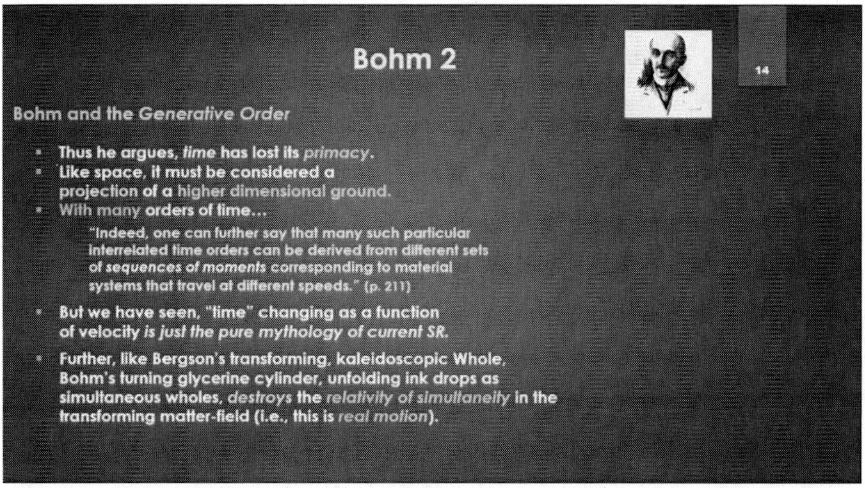

Steve:
Thus, he argues, time has lost its **primacy**. Like space, it

---

[1] To my knowledge, this is not a model that Bohm ever utilized, but I could be mistaken. Steve himself did use this model in his #10 – on Quantum Mechanics – particular in relation to Kastner's TI (Transactional Interpretation of QM) when discussing "flatland."

must be considered a **projection** of a **higher dimensional ground** – with many orders of time. The orders of time came from that previous quote because of Special Relativity. According to Bohm:

Indeed, one can further say that many such particular **interrelated time orders** can be derived from different sets of **sequences of moments** corresponding to [wait for it . . .] material **systems that travel at different speeds**. (p.211)

Pryzdia:
All right, so . . .

Steve:
This whole thing is coming from Special Relativity's bull crap.

Pryzdia:
Okay, so you are saying that for Bohm "time has lost its primacy" because of his reliance on the Special Theory Relativity.

Steve:
Basically!

Pryzdia:
Interesting.

Steve:
And again, go back to that model, the sphere -- I say it has lost coherence. Time has lost its primacy, but then what's projecting the freaking sphere through the flat plane?

Pryzdia:
Yes, I guess I had hoped that his reliance on the Special Theory of Relativity wasn't that fundamental. But it is fundamentally important as you lay it out here.

Steve:
So, as we have seen, "time" changing as a function of velocity **is just the pure mythology of current SR**. Further, like Bergson's transforming, kaleidoscopic Whole, Bohm's turning glycerin cylinder, unfolding ink drops as simultaneous wholes, **destroys** the **relativity of simultaneity** in the transforming matter-field (i.e., this is **real motion**). There are multiple ink drops coming out of the glycerin cylinder or could be -- all simultaneously. Well, they're clearly simultaneous. So, what happened to the relativity of simultaneity that he has to love and accept from Special Relativity in his own model. Again, coherence falls apart. To go on, Bohm states:

> Similarly, we are led to propose that this multidimensional reality may project into many **orders of sequences of moments** in consciousness. [Again, all this comes back to the Special Relativity thing.] Not only do we have in mind the relativity of psychological time . . . (p.211)

But further: "The fundamental law then is that of the immense multidimensional ground; the projections of this ground determine what time orders there may be (p.211)."

I say this whole thing is incoherent. I just can't get there. Somehow, time -- the transformation of the universal field -- is projected from orders (of "**sequences of moments**") in a multi-dimensional space (referring to the slide above).

Pryzdia:
Notice the reliance on space over time -- as opposed to Bergson.

Steve:
Right. And yet on analysis it crumbles because he has to introduce time somehow just to get that projection. I go back

to Bergson's quote -- everything ultimately resolves to motion, that is, to time. But, in my opinion, this is all predicated – these "multiple orders of time" -- on the wrong interpretation of SR. This is what I am complaining about – it stops me from the start. The "spatial order" has to be in motion (just like the ball!) – motion (thus time) is still primary! And what the heck put together that spatial order of the ball in the first place? I don't know. You see . . . the problem I have with some of the Bohm stuff is that it sounds beautiful, but that's the problem – it sounds beautiful until you . . .

Pryzdia:
"Look under the hood." [One of Steve's most common sayings.]

Steve:
Yes, open the hood a bit more and start thinking about trying to make sense of it. And because Bohm has such prestige, and he should have, nevertheless, that kind of hurts you a bit psychologically from opening the hood. You know? You don't want to question the salesman who is attempting to sell you a wonderful Mercedes or impugn his integrity by thinking that maybe there's 42 chipmunks under the hood pulling the rubber band -- twisting the rubber band. [laughter] That's my metaphor. So, let's go on.

Pryzdia:
Well, just for completion's sake, can you go back to the whole fish tank thing? You didn't make a direct reference to the graphic/illustration that you had in your slide, unless you're going to do that now.

Steve:
Yes, sorry about that. I got . . .

Pryzdia:
That's a key component of his model -- the fish tank metaphor. I think everybody who's familiar with Bohm knows how he uses it . . .

Steve:
Right. I mean, again, it looks like the fish isn't correlated. It really comes with that higher dimensional, three-dimensional fish tank. We're taking two orthogonal views of the fish tank with the two cameras. Notice even again, the incoherence there, because that three-dimensional fish tank has got all kinds of motion and time in it, doesn't it?

Pryzdia:
Right.

Steve:
It's hardly a static spatial order, is it?

Pryzdia:
It is not just a snapshot, that's right. It's a video, not a picture.

Steve:
And again, you can translate the same thing with regard to what the fish tank illustration is saying to the flat worm illustration just above it. If the flat worm focuses on two points on that black circle as they expand, he's wondering why they seem to spread out simultaneously. You know? Just like focusing on the fish – you wonder why its movement is precisely correlated as it turns toward you or away from you. So yes, same difference. And it's got the same problem. It is hardly a static spatial order.

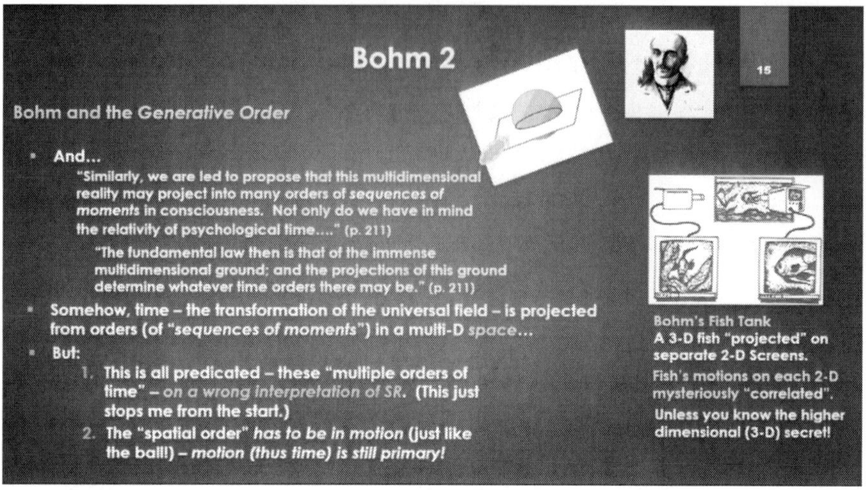

So again, my argument would be, as Bergson argued, **motion (thus time) is still primary!** In my mind, Bohm is basically working with the classic metaphysic of space and time.

Pryzdia:
He puts more emphasis on space than time.

Steve:
Yep. For Bergson, this classic metaphysic is a mental diagram, a "mesh" that we place under the concrete extensity of the external world.

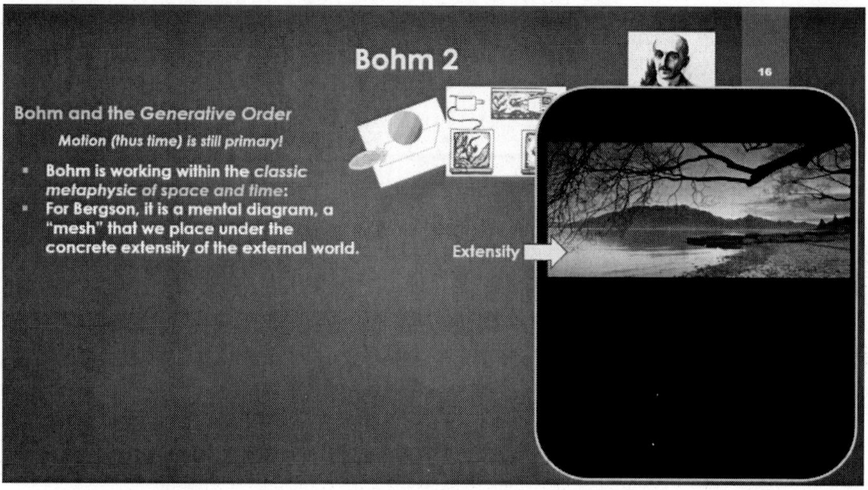

And by extensity we are looking at . . . there's a nice picture of extensity (see above) -- that lake surface and the shore, et cetera, et cetera. And it's before we place any conceptual framework of space over it.

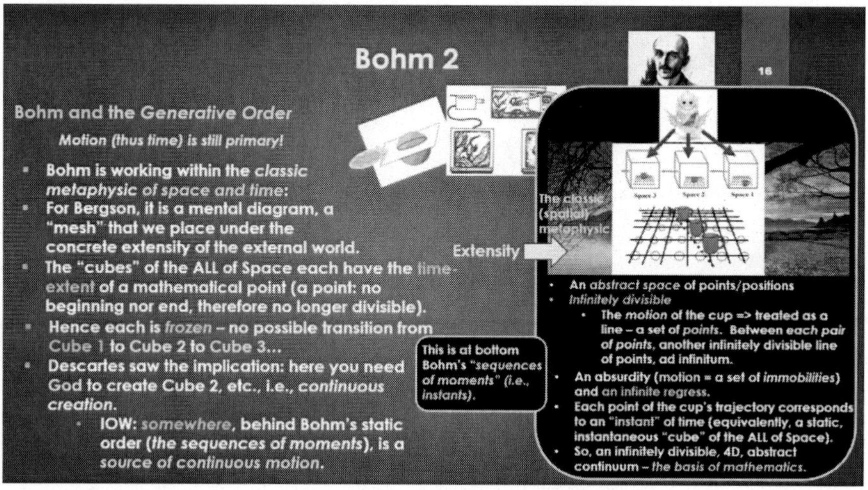

That kind of a framework (see above) -- where we place over it, as Bergson said, a kind of a mesh, a mental diagram. We can make those squares as small as you will down to the mathematical point -- each square can become a mathematical point. In fact, what it is an **abstract space** -- ultimately of points/positions that are infinitely divisible according to Bergson. So, the **motion** of the cup through that space is treated as a line – as a set of **points**. But between each pair of points, then there's another infinitely divisible line of points, ad infinitum.

So, in this treatment of motion via this abstract space of points and positions, you have an absurdity. Because motion now becomes a set of immobilities -- and it's **an infinite regress** -- because between each pair of points I can make another line of points; and again, between each pair of those points on that line, another line, et cetera, et cetera -- until I get to a mathematical point. A mathematical point is my stopping point because such a point has no beginning nor end. So, it is not divisible anymore. So as that cup moves, each point of cup's trajectory corresponds to an "**instant**" of time. In other words, you're building up the fourth dimension of this abstract space. For every point of that cup's trajectory, there is the equivalent of the ALL of space. We'll visualize it as a cube. It's an infinitely divisible, four- dimensional, abstract continuum – **this is the basis of all mathematics**.

The "cubes" of ALL of space each have the time-extent of a mathematical point -- a point being no longer divisible -- hence each cube is **frozen**. There's no possible transition from Cube 1 to Cube 2 to Cube 3 . . . This is at the bottom of Bohm's **"sequences of moments" (i.e., instants)** – this metaphysic. And note that Descartes, as Bergson noted, saw the implication: here you need **God** to create Cube 2 once you have Cube 1 – you need to introduce a God to create each cube after each cube -- because each cube is frozen -- you can't get to the next cube. Each cube at that mathematical point has no longer the possibility of changing to the next cube. You have to bring in a force to create the change -- which is precisely kind of what we're talking about with Bohm's problem. What is creating the motion of that sphere through the plane that the flat worm is observing? In other words, **somewhere** behind Bohm's static order (the sequences of moments), is a **source of continuous motion**.

Pryzdia:

Yes. This is interesting. I've always looked at the implicate order as being the "glue" that would glue together each of those cubes. But you're making the point or asking, "Where's the motion?" "Where's the movement coming from?" Correct?

Steve:

Right. Those series of cubes: Space [Cube] 1; Space 2; Space 3 . . . put them together in your mind . . . they're truly three dimensional . . . there'd be one giant four-dimensional structure, frozen, static, no motion whatsoever. And that's of course the problem with the space-time block of Special Relativity ultimately. You say, that's nice, you have a 4D block, but how do you get motion? Now you have . . . everything is an illusion. . . if I'm stirring the coffee, perceiving that motion, that's an illusion because actually it's all a 4D extended block where there is no motion. Same problem. He has the same problem. Once you accept the 4D space-time block of relativity, which is essentially what my little picture there is with the three spaces, put them all together in your mind and you have a fourth dimension. Now you have the fundamental question of a theory of the illusion of motion, the experience of motion. How in hell do you account for that now? Because the brain is just a four-dimensional object in that same structure. You have to have something like a consciousness field moving over those blocks to create the motion. But then of course you have to explain how the consciousness field registers the motion, et cetera, et cetera.

Pryzdia:
Okay. So that's the part that needs to be made clear. I think a lot of people can see his implicate order as being the glue that glues the cubes together. And then he speaks of the **generative order of the holomovement**, which **is movement**. But you're saying . . . be a little more clear on the whole motion thing because I think Bohm people will say, well yeah, there's the holomovement -- there's the generative order.

Steve:
Where does that come from though? Where does the holomovement . . . you've got one vast 4D structure in Bohm's theory . . . it's frozen. You need Descartes' God to make it move somehow.

Pryzdia:
That's fascinating. Because I think he would say that it is not a static hologram – it is not like a static photograph. I always looked at his theory as saying something to the effect that the universe can be metaphorically perceived to be a **moving hologram**. And that's why he refers to the holomovement. But your point . . .

Steve:
Yes. You and I are back to the unrecorded questions, the unrecorded conversation that we had with your profound questions that you asked me earlier.[2] How do you characterize duration as memory plus novelty/creation? You know? We're back to that -- evaluating that statement. Right here.

---

[2] See the opening dialogue of Talk 3 above.

Pryzdia:
Right.

Steve:
Okay. Let's see what's next here. [laughter]. It's all becoming a surprise to me.

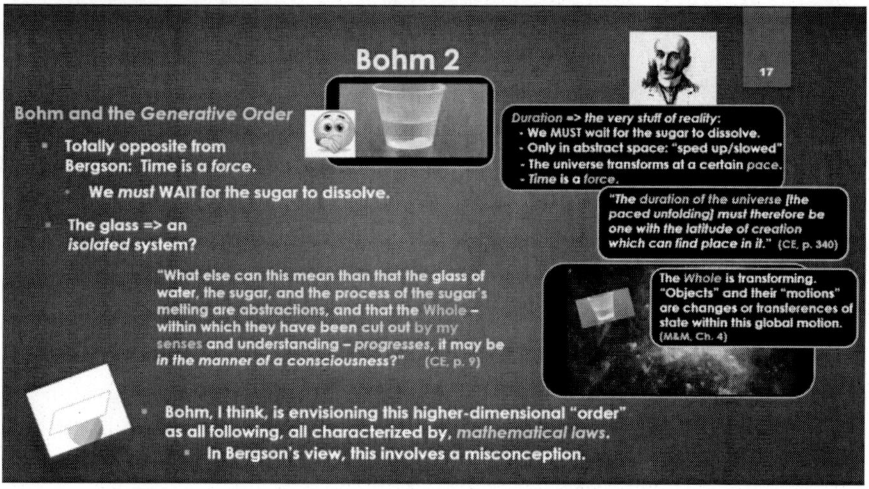

Okay, so here we go. **Totally opposite from Bergson where time is a force**. As Bergson noted: we must WAIT for the sugar to dissolve. Why do we have to wait? Because **duration is the very stuff of reality**. We must wait for the sugar to dissolve (see above). Only in an abstract space can you "speed up" or "slow down" the event at will. Because there's nothing inherently generating . . . If you think of that abstract space as my series of cubes all laid out, I can transition from one cube to another cube as fast as I want to, or as slow as I want to, because there's no concrete thing in my theory moving the cube from one to the next to the next to the next. It's just an abstraction. I can speed it up or slow it down at will. And of course, I don't even know what it is that I'm speeding up or slowing down -- because there's a purely mental thing that . . . oh yeah, there must be a Cube 1, then there has to be Cube 2, then there has to be Cube 3. But

I have no idea why, because in my abstract theory, it just all sits there as a series of static cubes. So, Bergson says, the universe transforms, the universal field, transforms at a certain pace -- and time is a force. And that is why you have to wait for the sugar to dissolve. And he says, “The **duration of the universe** [the paced unfolding of the universe] must therefore be one with the latitude of **creation** which can find place in it (*Creative Evolution*, p.340).” So that’s kind of back to that other comment – that quote of his – time is this hesitation before the next possibility, the next possible state of the universe, or it’s nothing. It’s almost a hesitation before . . . we’re trying to . . . what’s the next creative moment going to be? And that only moves so fast. I can only dissolve that sugar so fast. And that system, that little glass, is an arbitrarily demarcated system. The glass, the sugar, the water, all are such because of their relations with the whole external field, the whole field, the whole universe. So, I am extending this concept to the whole -- the whole universe is unfolding at a certain pace of which the glass and the water is a subfield, a subset, of that whole field unfolding at a certain pace.

Pryzdia:
Okay, can I come in here, Steve? Because I think we should clarify some things. Before we hit the “Record” button and started this formal presentation, you and I were talking informally about the “time is a force” notion. And I was expressing to you how I’ve always struggled with Bergson’s notion that time itself is creative because I’ve been so impacted by Bohm’s work who was himself influenced by Jiddu Krishnamurti; and these guys would make a distinction between “invention” and “creation.” Invention took place in time where something exists and you move it around and something new is revealed. But that’s not creation. Creation

needs to come from -- to use a term that Bergson despises: "nothingness." Or Bohm and Krishnamurti spoke of a "moving stillness" (or a "stillness in movement") -- from which creation can be generated. My point though is that with Bohm you have this other which "generates" something else. Thus, creation for me makes sense in Bohm because you have this "generative order" notion. And I have expressed this to you . . .

Steve:
But hiding behind that notion there has to be motion itself.

Pryzdia:
I just wanted to say though that I have tried to look at Bergson's "duration" as a Bohmian "generative order." And I think I mentioned that to you once and you didn't really respond to it. But I've had in the back of my mind that **why can't we look at Bergson's duration as a generative order?** Because it does "unfold" -- even matter itself. What do you think of that notion? [long pause]

Steve:
You have to be careful.

Pryzdia:
I assume so.

Steve:
Because you can't imply that "it's all there" in that order and it's just unfolding.

Pryzdia:
In the generative order?

Steve:
That's totally non-Bergsonian. Bergson would say, "that's just teleology." You're just unfolding toward an already defined goal. And that goal isn't defined in reality.

Pryzdia:
And Bohm does flirt with teleology. He really does. He does adhere to Aristotle a lot. Well, alright – let's just say for now that it is a "subtle issue."

Steve:
Remember, we are talking about "becoming" here; and Bergson rejects both mechanism and finalism (teleology). Mechanism is just a mechanical unfolding – Laplacian, determinist, mechanical unfolding. And finalism is just the same form of mechanical unfolding, all being drawn to an already specified goal or vision. For instance, I have a mousetrap in mind, and I'm unfolding all of the events (putting together my box and my pencil and my rubber bands, et cetera) all unfolding toward my ultimate mouse trap that looks like a crossbow.[3] And he would reject that too, because neither of those actually affirms the creativity that comes out of duration. Yep. We're on a tough subject here. [laughter]

Pryzdia:
To say the least.

Steve:

Alright, so shall we go on here?

Pryzdia: Sure.

Steve:
Now, as I have preached already, the glass is an isolated system. As Bergson says:

---

[3] See Steve's books and videos on "Meditations on a Mousetrap" – the Mousetrap and the Complexity of Devices (with regard to "Evolution and Mind").

> What else can this mean than that the glass of water, the sugar, and the process of the sugar's melting, are **abstractions**, and that the Whole -- within which they have been **cut out** by my senses and understanding -- **progresses**, it may be in the manner of a consciousness? (*Creative Evolution*, p. 9)

The whole being the whole universal field is progressing at a certain rate. The very stuff of reality -- time is a force. In other words, the glass is just part of the transformation of the universal field. And then you can go back to what he said in *Matter and Memory*: The Whole is transforming. "Objects" and their "motions" are changes or transferences of state within this global motion. (*Matter and Memory*, Chapter 4) And that's already where you come to the simultaneity of flows that he later talked about in 1922 in *Duration and Simultaneity*. Because the organic growth of the rose is in reality just changes or transference of state within this global motion. It is a simultaneous sub-motion within the global motion -- along with Einstein's thunderstorm. It's interesting: all the Bergsonians, everybody I've seen commenting -- they don't recognize that they don't grasp how Bergson's comments in *Duration and Simultaneity* about simultaneity and of flows was already there in *Matter in Memory* in 1896. In fact, I think he said it better in *Matter and Memory* than in *Duration and Simultaneity* -- which I can understand because you forget what you said after it's . . . 25 years later. [laughter]

Pryzdia:
Yeah.

Steve:
So, Bohm I think is envisioning this higher-dimensional "order" as following, as all characterized by, **mathematical**

laws. But in Bergson's view, this too involves a misconception. See, I think if you ask: "Does all this unfold, Dr. Bohm by mathematical laws?" – he would say yes. It all unfolds according to the laws of physics in my classic metaphysical framework of space and time, abstract space. It would have to.

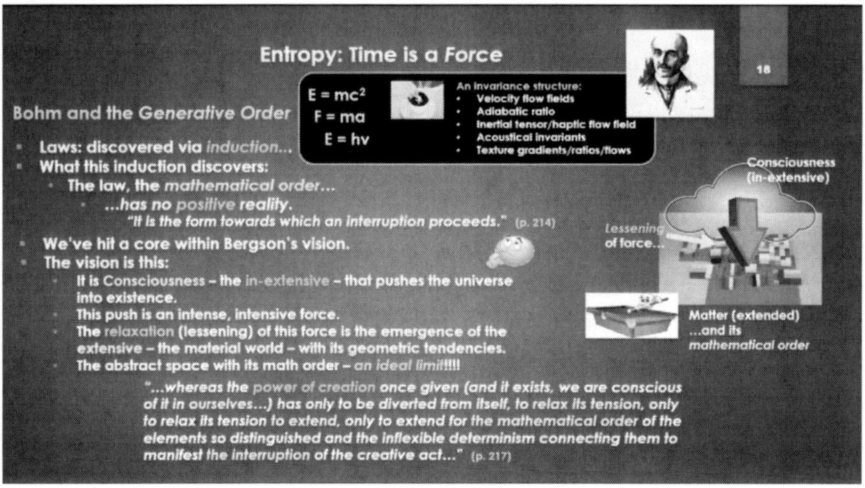

What's the misconception? Well, for Bergson, these kinds of laws are discovered by **induction** – because they are called **invariance laws**. So, if I do enough experiments and stir the coffee enough it always makes the radial flow field – that is an invariance law associated with coffee stirring – it is an inductive process. But what he says is that what this induction discovers is that the law, the **mathematical order has no positive reality**. What does that mean? Well, according to Bergson, "It is the form toward which the interruption proceeds." So then we've hit this core of Bergson's vision again -- back to our becoming. The vision is this: It is **Consciousness** – the **in-extensive** – that pushes the universe into existence.

So, I have my picture there (see above) of the in-extensive consciousness, but it's pushing, thrusting the universe into existence. But this push is an intense, intensive force. As it proceeds toward the world of matter, there is a relaxation (lessening) of this force, and this relaxation is the emergence of the **extensive** or the material world -- with its tendencies. But the ultimate expression of this as a precise mathematical order treated geometrically in the abstract space with its points and positions of mathematical instants – that is **an ideal limit**!!! That that arrow never quite pushes everything to the mathematical order that a physicist like Bohm would envision actually exist. And, of course, that is where all of science is making a mistake right now -- they've exalted this **mathematical order**, which is an endpoint and **an idea limit** into the point in which you **start**. And it's not the case. It is just like artificial intelligence – this is also simply an expression of the classic metaphysic. The Turing Machine is this state to state, instant to instant, model with its abstract manipulation of spatial objects -- which they think is where you **start** a theory of consciousness. But it's **not** the point. That abstract space is a derivative of cognitive development of the brain. But the brain doesn't exist in that abstract space. It has created the abstract space. It exists in the **temporal metaphysic**. **And time is a force**. And so everyone's starting at the wrong point. They are starting from the end point, which is basically an abstract limit derived by the brain. And that is screwing everything up. This is my point: if you were to ask Bohm "Well is this all unfolding by a mathematical order?" He would probably say yes – and Bergson would say no.

Before I read the following quote, I am going to expand on it. Because Bergson gives us an example of finding an inductive law -- boiling water. Okay. You boil a pot of water, then you repeat the process the next day, then you repeat the process the next day. You are basically super-imposing upon an ever-changing universe an abstract situation where you're pretending that everything is the same from Day 1 to Day 2 to Day 3 -- you have the same pot, the water is basically the same, you have the same heat . . . And you're coming up with a law of boiling. And Bergson says, nature never dreamed of such a process; which is to say, it never dreamed of such a law. So again [laughter] -- nature never dreamed of such a process. You have to sort of let that one . . . that's in *Creative Evolution*. These laws have no positive reality -- but physics sure thinks they do. And I think Bohm sure thinks they do. So, Bergson says in *Creative Evolution*:

> . . .whereas the **power of creation** once given (and it exists, we are conscious of it ourselves . . .) has only to be diverted from itself, to **relax its tension**, only to relax its tension **to extend**, only to extend for **the mathematical order** of the elements so distinguished and the inflexible determinism connecting them to manifest **the interruption** of the creative act . . . (p.217)

I mean we have to go back to our original discussion, to what we talked about before we hit the "Record" button today.[4] We have to ponder all of this in the context of your question. But anyhow . . . I think that's all I have to say here on that. The question is: Does this generative order really have the positive reality that you think it does? I don't think so.

Pryzdia:
Like we said, it's very subtle. We need to continue a discussion around this topic. I just want to go back to that last slide because, like we said, before we officially started this presentation, we were talking about how Bohm speaks of enfolding and unfolding from a generative order and back

---

[4] Again, see the opening dialogue of Talk 3 above.

and forth. And yet Bergson speaks of a tension and extension, but it is all of one order. It's really very subtle -- how this is put out there.

I think at some point I would like to discuss, obviously not during this presentation, but I think it would be worthwhile to bring in Jiddu Krishnamurti's influence on Bohm. For instance, I'll just throw this out there: I was watching the 1975 Krishnamurti/Bohm Dialogues. The transcripts now exist. These particular dialogues have never been published as a complete set. But I was watching Dialogue #10 yesterday afternoon, and they were talking about creativity, which ties into what we're talking about here. And Bohm makes the following point: he states that **creativity itself is not created**. And they start talking about this, and Bohm says something like **perhaps even in nature, the depth which brings about something is not of time**. **The creation of new forms in nature may also depend on what's beyond time**. So, this whole notion of a "realm that is still but yet in movement," which is "generating creation itself," is again a very subtle thing.

Steve:
Well, we've got a schematization of the problem right there with my arrow going up and down (see above). What is that in-extensive consciousness that is generating the external world? You see, it has vague resonances to Bohm. Right? To, as you say, a generative order. But it's not going to be the generative order that Bohm might be thinking. You know? But yet it's something . . . when you say in-extensive, it's not exactly what we consider . . . well, this is another whole interesting question.

Pryzdia:
It is very subtle.

Steve:
Yes. Is there really the "in-extensive" in Bergson? And the reality is that he started out that way in *Time and Free Will*, but then modified that to everything had at least some extension.

Pryzdia:
That's a very important notion.

Steve:
Right. So even that in-extensive thing is kind of . . . you have to think about that. Although that is what he has in *Creative Evolution* – that's the picture. And I believe he uses the term in-extensive. So . . . yes, you could see . . . I mean that little picture there, it sort of looks the same as Bohm's – he didn't picture it this way, but the ball going through the two-dimensional plane . . . but yet, it is different. I think if we go to the next slide . . . I think I have one more slide on the generative order.

Pryzdia:
Okay.

Steve:
Should we do that?

Pryzdia:
Yes.

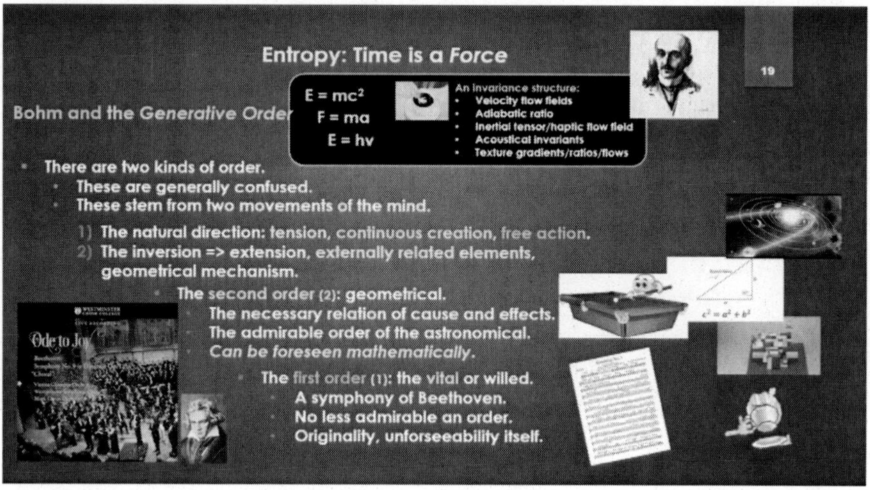

Steve:

Bergson says that there are two kinds of order. These are generally confused. These stem from . . .this is interesting . . . these stem from two movements of the mind. The natural direction is **tension, continuous creation, free action**. The inversion is **extension, externally related elements**, geometrical mechanism. In other words, that abstract space I described. The second order is geometrical. It's a necessary relation of cause and effects -- the admirable order of the astronomical, all the planets work in their nice little orbits or whatever. And all **can be foreseen mathematically**. The first order is the vital or the willed. A symphony of Beethoven. No less admirable an order. Originality, unforseeablity itself. And so, if you went back to my little picture there of the extensive cloud, in that would be this first order, the vital or the willed. But it's not the same kind of order as the order of mathematical law that fits in the abstract spatial continuum of the classic metaphysic. So, if you were to say, well, Dr. Bohm, if you were to actually . . . this is what is so frustrating because some of Bohm's discussion about music et cetera almost gets to this. Like . . . well, really, we're talking about the first order, the vital or the willed, the symphony of Beethoven -- that is the generative order that ultimately generates the whole series of notes that Beethoven wrote out . . . whatever it was . . . the hundred eight pages that it took to write his symphony.

Pryzdia:
But notice our language here Steve I just want to note the use of the term “order” -- one order, another order.

Steve:
Right. Right. Right.

Pryzdia:
I’ve always felt that for Bohm order is more fundamental than time or space. And for Bergson, time is more fundamental than order. Would that be a correct . . .

Steve:
Right. And for Bergson you have to have the time to create the order. I think that’s going on.

Pryzdia:
That’s what is going on.

Steve:
That’s why it’s so frustrating – Bohm gets so frustrating resonant with something in Bergson.

Pryzdia:
I think anybody who has studied Bohm deeply will agree with what I’m about to say. He really struggled with the nature of time. I give him credit though, anybody would. He really did a lot of meditating on the nature of time. And he did a lot of this with Jiddu Krishnamurti. He did see “the problem of time” – but for him it was more of a paradox than a problem. He admitted . . . he said no one has understood the nature of time in his opinion. But he did place more of an emphasis on the concept of “order” – for him this was more fundamental.

Steve:
Well, yes. I think we are really getting down to the fundamental similarity and the fundamental difference all simultaneously here with him. I think we're getting there as we compare and contrast him to Bergson. By the way, the reason I have that little guy with a coffee cup there down the bottom right (see above), is because that's voluntary action. Even that -- just reaching out for that coffee cup and taking a drink is coming from the willed or vital order. In other words, our everyday actions, as long as they're not automatic/robotic, are coming from this deep inner vital or willed order.

Pryzdia:
It's an important point.

Steve:
Which leads to the problem of voluntary action and free will. But that's another interesting question. Let's peek ahead to the next slide. Because I think you said that you want to maybe stop . . .

Now we're going to get to this topic: "Pulses, stable sub-totalities, Whitehead, Capek, etc." Let me . . .

Pryzdia:
OK. Wow. Well, we have been going for an hour and 45 minutes. But this topic is important to me. Do you want to move forward or do you want to wait for another day on this? Because this to me is crucially important. The Whitehead, Capek, Bohm, Bergson thing. How much do you have on this? Do you have a lot?

Steve:
I'm going through them right now. Slide 20. Slide 21. Slide 22. Three slides. But probably heavy. Then I go to Susskind

and the Holographic Principle – a couple slides -- two, three slides there. Oh my God. Even more because we . . .

Pryzdia:
These are important slides.

Steve:
Yep. I got a lot on that too. Maybe we better do another . . . yes, I'm probably running out of steam.

Pryzdia:
But this is good.

Steve:
There are probably five or six on the Susskind's Holographic Principle.

Pryzdia:
But this to me . . . what we're about to hit -- the pulses and then the Holographic Principle of Susskind -- this is quite important to get clear on with regard to Bohm. So, I agree. I think we should do another presentation – and fairly quickly, by the way, so all this stays fresh in our head. My schedule's fully open. So, if you want to do . . .

Steve:
Okay, well we can just do this tomorrow.

Pryzdia:
I can do tomorrow. All right. Tomorrow at . . .10:00 or 9:00?

Steve:
At 10 again?

Pryzdia:
10 your time. Correct?

Steve:
Let's make it 10:30.

Pryzdia:
10:30.

Steve:
I had to rush to finish my coffee. [laughter]

Pryzdia:
Yes, that's fine.

Steve:
But 10:30 feels better because I'm a little fresher.

Pryzdia:
Okay.

Steve:

Oh, and I have a final slide here on reconciling Bohm and Bergson, which I probably should look at before we talk. But yes. So basically, I have two more topics -- the pulses topic and the Holographic Principle topic. There's still like ten slides here, I think.

Pryzdia:
So. And then you have a conclusion.

Steve:
Yes -- on reconciling Bohm and Bergson – but I think we actually did a better job on than this slide does. [laughter] But I'll look at that slide again. Good, good. So, this is a good thing -- because I think we just saw one point where you'd reconcile Bohm and Bergson and that would be in this notion of the in-extensive in Bergson, the willed or vital order that requires time to even come about.

Pryzdia:
That is a very important notion to meditate on with regard to Bergson and Bohm.

Steve:
Because to put it simply, it's back to that notion that for Bergson time was a reality and for Bohm it wasn't.

Pryzdia:
It wasn't a fundamental portion of his vision. But then again, I'll tell you what, that's going to tie into what you're going to do tomorrow -- the whole pulsation thing. So that's important, and we will come back to this content tomorrow. I'm sure.

Talk 4: On Bergson and Bohm, Part 2

Pryzdia: Okay, Steve, we are back and I am looking forward to . . . I guess we're calling this Part 2 of Bohm 2. We're going through a Bohm-Bergson comparison/integration, and we got a lot done in Part 1 -- moving through very important notions. So now we'll just pick it up when we left off and you can just take it away.

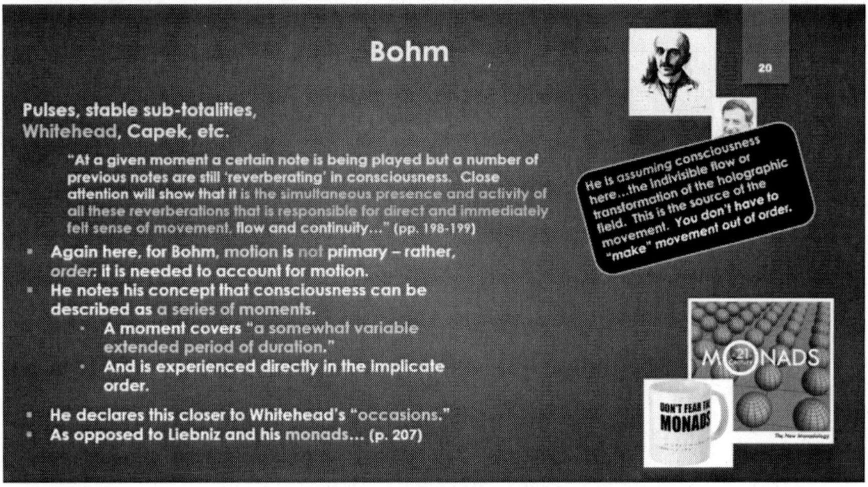

Steve:

We're picking it up from the previous discussion on "Bohm and the Generative Order" versus Bergson. And so now we are onto another topic that you wanted to take a look at -- namely pulse's, stable sub-totalities (in Bohm's framework or phraseology), Whitehead (we're going to be looking at epochs), Capek's, notion of pulses, et cetera. I have a nice quote here from Bohm:

> At a given moment a certain note is being played, but a number of previous notes are still 'reverberating' in consciousness. Close attention will show that it is the simultaneous presence and activity of all these

> reverberations that is responsible for direct and immediately felt sense of movement, flow and continuity . . . (*Wholeness and the Implicate Order*, pp. 198-199)

So that's from *Wholeness and the Implicate Order*. Now what I'm noting here is that he is **assuming** consciousness here, the indivisible flow or indivisible transformation of the holographic field. This is the source of the movement. **You don't have to "make" movement out of order**. OK? Which is what he's implying here.

Pryzdia:
OK . . .

Steve:
So, first comment: Again here, for Bohm, **motion** is **not** primary – rather, it's **order**: it is needed to account for motion. He notes – and here we are just continuing on from the previous subject discussed in Part 1 -- he notes his concept that consciousness can be described as **a series of moments**. A moment covers "a somewhat variable extended period of duration." So, the moment can be of various extents in time -- but we've got the moments -- and is experienced directly in the implicate order -- which is an interesting statement because it is experienced directly in the implicate order, but we only experience or at least see, the explicate order. So, I don't quite understand that statement. But he says it. He declares this is closer to Whitehead's "occasions," as opposed to Leibniz and his monads (p.207). I'm not even going to make a comment on that because, now that I look at it and the fact that I put it there, I don't know truly what he meant. Although I do understand what he means by it being close to Whitehead's occasions or epochs – variable moments of duration.

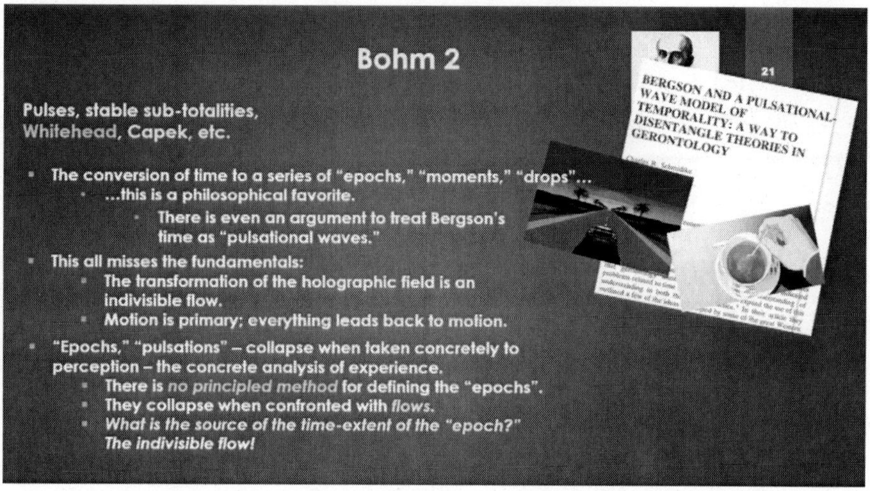

So, my comment: the conversion of time to a series of "epochs," "moments," "drops" -- this is a philosophical favorite. There's even an argument that you put me on that we need to treat Bergson's time, his duration, as pulsation waves; and I have this article here from Schmidtke on Bergson and the pulsation wave model of temporality ("Bergson and a Pulsational Wave Model of Temporality: A Way to Disentangle Theories in Gerontology"). But you have pointed out that this already started with Capek in at least 1971; thus, this is more from Capek than from Schmidtke.

But to me this all misses the fundamentals -- that the transformation of the holographic field is an indivisible flow. And to try to make that indivisible flow into drops or moments or epochs that jerk along . . . in the Whitehead version that I talked about in #53 of my video series it is particularly apparent that he has no method of gluing together the epochs -- you have Epoch 1, then you jerk to Epoch 2, then you jerk to Epoch 3. In other words, these drops or moments -- there's no way to account for how they are strung together. In fact, you're back to assuming the indivisibility of motion, where at best these things would be sort of, well, okay, pulses I guess -- in this indivisible motion where there's no real discontinuity between one and the next -- they bleed into each other. But do you really need them? I don't know. Again, I note: motion is primary; everything leads back to motion. The epochs, the pulsations, kind of collapse when taken concretely to perception -- that is to the concrete analysis of experience. There is **no principled method** for ever defining the "epochs." It always just is sort of made up ad hoc as some theorist decides to say, okay, well that's an epoch and that's an epoch. Just looking at the coffee cup (see above): Is each 360 degrees circle of that spoon an epoch? One circle is an epoch? Two circles are an epoch? What is an epoch in the coffee stirring? You know? Or is the whole coffee stirring an epoch? Which . . . then the epoch could be . . . if I choose to stir for ten minutes, is that a ten-minute long epoch?

You see where it starts to break down? Or look at that flow field (see above) as I drive down the road . . . what's an epoch there? As each tree goes by -- is that an epoch? What if there are no trees? I mean what if it's just a desert road -- like in Arizona -- where would the epochs be in that? And if you drove for two hours, is that a two- hour long epoch? You see my point? There is no principled method of doing these epochs, and it sort of collapses when you look at Gibson's optical **flows** that characterize these events.

So, you're just struggling to try to slap this epoch notion on top of experience. And again, **what is the source of the time-extent of the "epoch" in the first place?** If my epoch is one 360-degree circle of that spoon, the source of the continuity of seeing that is still the **indivisible transformation of the universal field**. You see my struggle with epochs is that they are just too artificial. It's just an imposed structure that these philosophers seem to love but doesn't buy us anything. Again: What's the start end of an epoch (see above)? How long does it extend? One minute? Ten minutes? No idea.

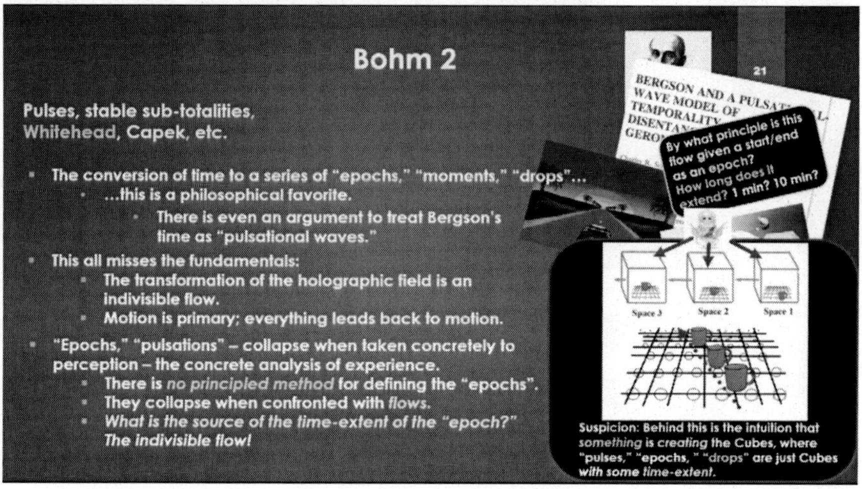

And my little note here (see above) is that the suspicion behind the epoch thing (or the moment thing) is this intuition that **something** is **creating** the cubes, where "pulses," "epochs," "drops" are just cubes with some **time-extent**. Remember we're back to the abstract space where each point along the motion of that cup is an instant in time, which corresponds then to an instant of the ALL of space -- which I'm visualizing as a cube.

So, each cube would be an epoch. You know? But then I have to have something that generates . . . something continuous like my God there (see above) . . . that generates the cubes -- with some time extent again. You see my difficulty with this Bohmian pulsation thing. It just doesn't buy much -- I guess that's the bottom line.

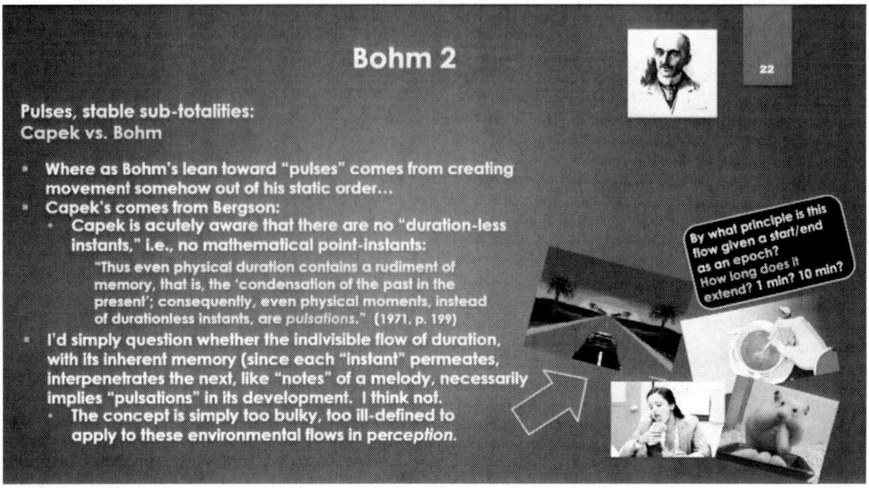

Whereas Bohm's leaning toward "pulses" comes from creating movement somehow out of his static order . . . I think Capek's comes from Bergson. Capek is acutely aware that there are no "duration-less instants." That is, there are no mathematical point-instants. So, he says:

> Thus, even physical duration contains a rudiment of memory, that is, the 'condensation of the past in the present'; consequently, even physical moments, instead of durationless instants, are pulsations. (1971, p. 199)

I'd simply question whether the indivisible flow of duration, this organic continuity, where each moment, as Bergson said, penetrates/permeates the next (and he is just using moment because you have to use something like that because of language -- because of how language is sort of based spatially) like the "notes" of a melody necessarily implies "pulsations" in its development.

Again, I think not. You see Capek, in his writings, as I told you yesterday, just sells this idea that pulsations are a necessary consequence of Bergson's duration. And I say, no, not even close. Again, notice, as I have said before, you're back to the notion that well, if it's so necessary, then why are you struggling to fit it to these flow fields? Why does it obviously become completely arbitrary and at ad hoc as to how you apply a series of pulsations to get flowing down the road there or stirring the coffee or eating the carrot (see above)? The concept is simply too bulky, too ill-defined to apply these to the environmental flows in perception. Okay. That was my comment there on . . .

Pryzdia:

Well, yes, let's summarize a bit. And I really want to thank you for clarifying that for me because as you know, this whole pulsation thing, whether it's through Capek or Bohm, it's been driving me nuts; because as I've said to you, most of the people I read who are critics of Bergson's work -- other than you -- I mean I think you're the purest -- because everybody else somehow or other goes back to those notions of pulsations, and they bring in and they try to integrate Bergson and Bohm, Whitehead. So, I really appreciate what you shared there. And again, I think we discovered something yesterday.

Let's kind of summarize a bit, because I think what we're learning is -- as I've always thought -- for Bohm, order is more fundamental than time; and for Bergson, time is more fundamental than order. And for Bergson, that order of time is in movement; and you are kind of saying, well, Bohm's notion of order is too static. And when you look at his work in quantum physics where he is trying to tie in the Special Theory of Relativity (STR), it becomes a little bit incoherent because he does have to account for movement. And for him -- he used to always say that order is very fundamental -- it is almost something that cannot be defined. But he did say that if order could be defined, he would say something like "order is perception." But this ties into his work with Jiddu Krishnamurti; they went on and on about how order is perception; we won't go into it now, but I just want to throw this out here because again, as I said to you before, I really think it's a sin that people who are looking at Bohm somehow or other try to avoid his interaction with Krishnamurti, which you just can't do.

Steve:
Well, that would be me. [laughter]

Pryzdia:
It's very common though. But I want to say . . .

Steve:
Well, I didn't even know about it until you told me.

Pryzdia:
Oh really?

Steve:
Yeah.

Pryzdia:
Well, let me just put this out there. When I first discovered your video series and you were talking about Bergson's notion of "pure perception," I hadn't read any Bergson yet, but I had read a lot of Krishnamurti and Bohm -- who used to use the phrase "pure perception." Now remember you and I were writing back and forth . . .. but Krishnamurti and Bohm's use of the term pure perception is entirely different than Bergson's use of the term. And I want to make the point that for Krishnamurti and Bohm . . . well for Bohm even order is perception. And Krishnamurti had this notion that **perception can act -- it has its own action**. This view is very influential in Bohm's work -- that order is perception -- and somehow for these guys, perception has its own action, which by the way, is what **they** called pure perception. That's

a perception without memory. So again, very similar to Bergson.

For Bergson, pure perception is perception without memory; and for Bohm and Krishnamurti, pure perception is perception without memory. But that **pure perception** for Bohm and Krishnamurti **has its own action**. Now, I'll leave it at that because I know we need to move on to other topics, but that's a very important issue for Bohm; but you can see there, if you view order as perception, if you think that that order as perception has its own action, now he's accounting for movement. But again, I think you would say that that's just a bit too abstract. There's no time scale there. There are no flow fields. I get that. I understand that.

Steve:
Right. I'd have to look at why they say that perception has no action.[5] It has vague resonances to Bergson's definition of per "perception as virtual action." Remember, for Bergson, perception is actually virtual action. I'm seeing how I can act on the holographic field in the future. As we've talked about, perception is actually the display of a past transformation of the universal field and simultaneously the display of how I can future act on the field -- virtual action. So, in that sense, perception would have its own action. Also, what I just said is equivalent to Gibson's "affordances" -- that I'm perceiving what the coffee cup and the spoon are affording in terms of action. But whether they mean that or not, I kind of doubt it.

Pryzdia:
They're coming at it from a more "mystical" perspective. This is Krishnamurti's notion.

---

[5] See above. For Jiddu Krishnamurti, perception IS action.

Steve:
Alright.

Pryzdia:
But what you just said there about the notion of virtual action -- that's very unique and very important and very powerful. That's just not in Bohm at all, at least not explicitly.

Steve:
True.

Pryzdia:
So, there are some subtle differences and some subtle similarities as well.

Steve:
Right. Okay. Interesting. Well, let's go to the next topic that you think is important, which I think is too, but I tend to ignore it.

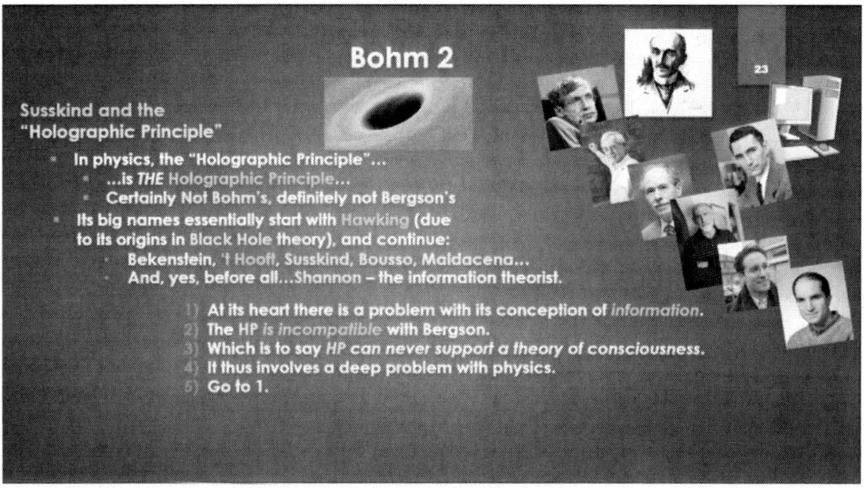

That is Susskind -- and it's not just Susskind – there are a whole list of people associated with the "Holographic Principle" of physics. And I'm noting here that in physics, the Holographic Principle is THE Holographic Principle -- but it is certainly not Bohm's, which is interesting, and not

Bergson's either. It's big names essentially start with Hawking (due to its origins in Black Hole theory), and continue with: Bekenstein, t'Hooft, Susskind, Bousso, Maldacena, and others. And yes, before all . . . Shannon – the information theorist -- because they all tie this back to the notion of information. There are five problems with the Holographic Principle.

1) At its heart there is a problem with its conception of **information**.

Pryzdia:
Bullseye.

Steve:
2) This Holographic Principle (henceforth HP) is incompatible with Bergson. Nor is it compatible with Bohm for that matter.

Pryzdia:
No, it is not.

Steve:
3) Which is to say the **HP can never support a theory of consciousness**.

4) It thus involves a deep problem with physics and then . . .

5) Go back 1 above [laughter] – at its heart there is a problem with information theory. And so, I'm going to spin this out a bit.

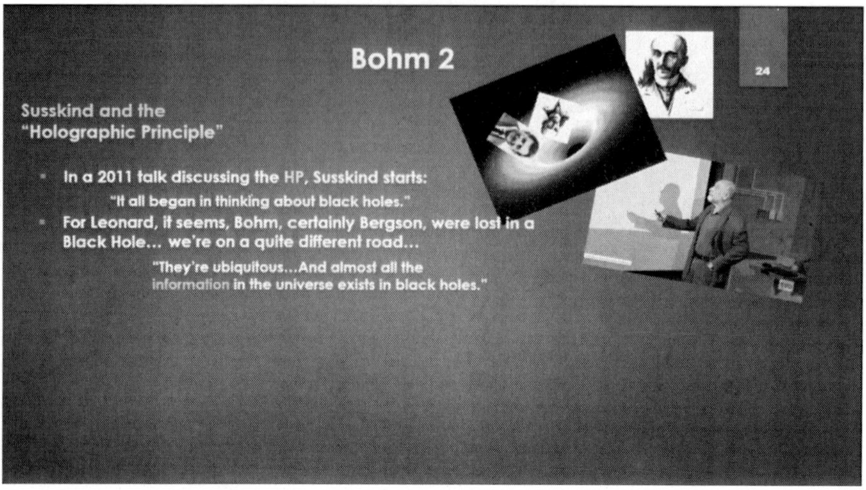

In a 2011 talk discussing the HP, Leonard Susskind starts with the following statement: "It all began in thinking about black holes." For Leonard, in other words, it seems Bohm, and certainly Bergson, were lost in a Black Hole. They don't exist. It's amazing to say the holographic principle starts with black holes. I mean that's an amazing piece of ignorance right there. You know? Unless of course you think the only holographic principle **is** their holographic principle. So, we're on a quite different road. He says: "They're [black holes] ubiquitous . . . And almost all the information in the universe exists in black holes." Amazing statement! What is information to them? Well, it is this . . .

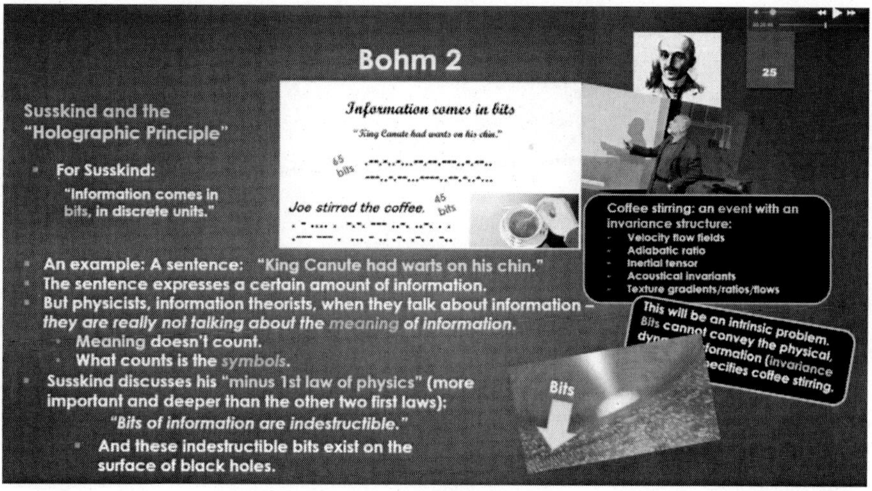

For Susskind: information comes in "bits." And he has the sentence there (see above): "King Canute had warts on his chin." And we can translate that to morse code there -- dots and dashes, and count them – there are 65 bits. Or I could have "Joe stirred the coffee." That would be 45 bits, which is describing that event there. Joe stirred the coffee – and you can already see the problem: if I'm viewing language as a modulating system to create the perception (internal/external) of an event, like coffee stirring, and that's the meaning -- you're going to reduce that to 45 bits? -- that stirring a coffee there? See, we've already got a fundamental problem. But for Susskind, "information comes in bits, in discrete units."

An example: A sentence: "King Canute had warts on his chin." The sentence expresses a certain amount of information. But physicists, information theorists, when they talk about information -- **they are really not talking about the meaning of information**. **Meaning** doesn't count. What counts is the **symbols**. This will be an intrinsic problem. Bits cannot convey the physical, dynamic information (invariance laws). There are invariance laws that describe or specify stirring coffee. I mentioned this before[6]: coffee stirring is an **event** with an **invariance structure**: velocity flow fields, adiabatic ratio, inertial tensor, acoustical invariants, texture gradients/ratios/flows. This just doesn't exist in the physics world -- which already tells you that they're basically cut off from reality.

Susskind discusses his "minus 1st law of physics" (he believes it to be more important and deeper than Newton's first two laws). He says, "Bits of information are indestructible." And these indestructible bits exist on the surface of black holes. Now this is an amazing statement which we're going to rip into and destroy in a second.

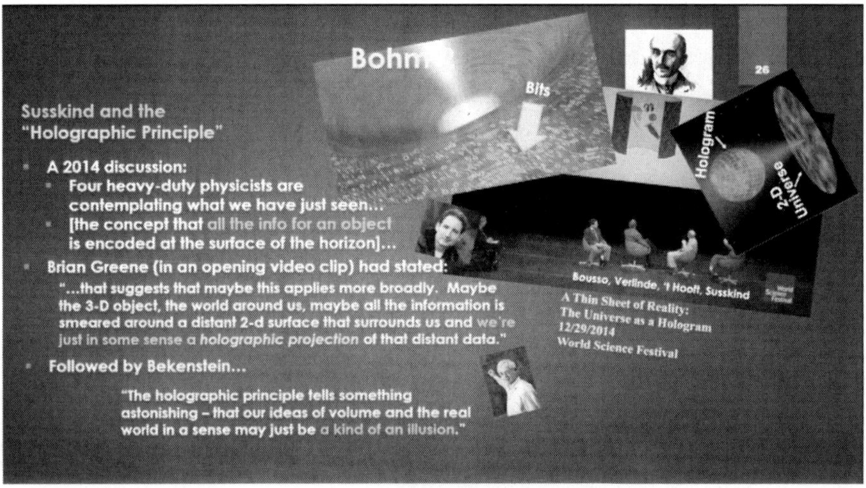

Fast forward to a 2014 discussion. We have four heavy-duty physicists to include Susskind there (see above). On a stage you have Bousso, Verlinde, t'Hooft, and Susskind. And they are contemplating what we've just seen Susskind talk about -- the concept that **all information for an object** is encoded at the surface of the horizon of a black hole. Back to

---

[6] See Talk 1: "On Bergson and J.J. Gibson."

the bits thing. And Brian Greene who is, I think he was . . . no, he's not the moderator there, but they played a video of him and he had stated:

> . . . that suggests maybe this applies more broadly. Maybe the 3D object, the world around us, maybe all the information is smeared around a distant 2D surface that surrounds us and **we're just in some sense a holographic projection** of that distant data.

So that picture there in the right corner . . . we're talking about a 2D surface that projects a 3D world, physical world. I mean your mind boggles when you think of how you're going to have . . . what that transition implies -- from the bits to a 3D physical world of stirring the coffee. Now that's followed by Bekenstein who says: "The holographic principle tells something astonishing – that our ideas of volume in the real world in a sense may just be **a kind of an illusion**." Now this is because everything about this holographic principle is two-dimensional, a 2D flat surface. And, of course, we live in a 3D, actually a 4D world. So, if everything's just a 2D surface, well then, the 4D world is an illusion. They have no theory of the illusion by the way, but that doesn't mean anything to them either. They're basically now theorists of consciousness in a subject where they actually don't really belong.

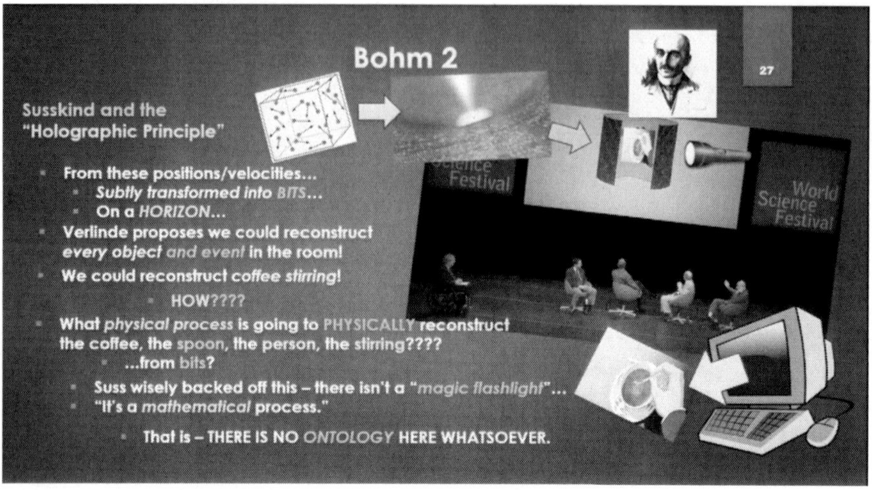

They start talking about the velocity and positions of particles, and they subtly transform these things into BITS on the HORIZON of a black hole. And then Verlinde proposes that we could reconstruct every object and event in the room! Interesting. We are back to my statement – you have a flat horizon or flat surface, which you're now going to construct the whole four-dimensional concrete physical reality. They think they can do this.

An example up there (see above): notice they're basically shining a reconstructive wave symbolized by my flashlight through that flat film surface and reconstructing the image, 3D-image, so to speak, of a clown apparently smoking a cigar. So that's their model. Somehow, we transition from 2D informational bits to a four- dimensional concrete world. In other words, we could reconstruct the coffee stirring. How? Well, they never approached this. What physical process is going to physically reconstruct the coffee and the spoon the person is stirring from bits?

Well, Susskind wisely backed off of this. He says, well, there isn't a magic flashlight. It's a mathematical process. Notice that. In other words, there is no ontology here, no actual physicality whatsoever. Somehow, they're creating all the reality from abstract mathematics. I mean the absurdity is gob smacking, but this doesn't stop them.

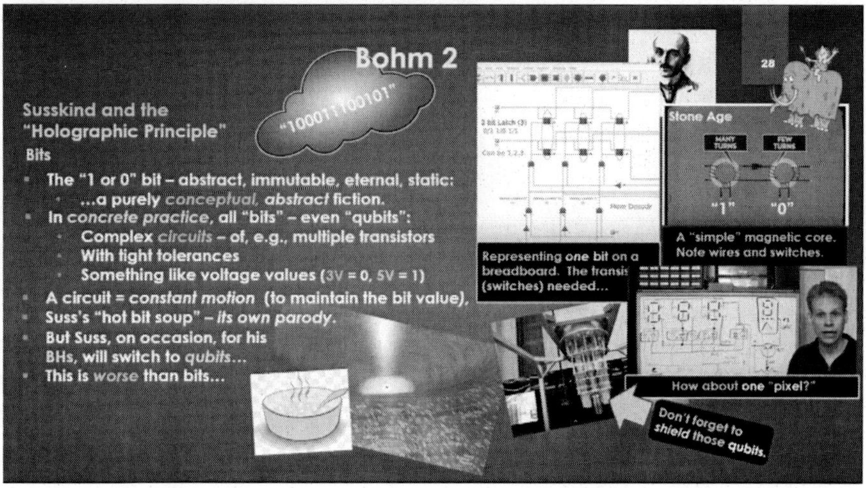

So, let's look at bits. The "1 or 0" bit -- abstract, immutable, eternal, static. Remember "the minus first law" of Susskind: indestructible. This is a purely **conceptual abstract** fiction. In concrete practice, all bits, even qubits, are complex **circuits** -- circuits of multiple transistors with tight tolerances – something like voltage values (say 3 volts = 0, 5 volts = 1). God help you if you have four and a half volts -- it's not going to equal 1 or 0. There's a picture (see above) of representing one bit on a breadboard -- all the transistors and switches needed to represent a bit.

Or here's the stone age version (see above), back in the old magnetic cores. Note all the wires and the switches in the turn. That's a bit. According to Susskind, bits just sit on black holes. What is he talking about? Here's a pixel (see above). He also talks about pixels. Well look at the picture on the board just to represent one pixel. OK?

A circuit is in constant motion (to maintain the bit value). So, Susskind's "hot bit soup" is **its own parody**. It's an absolute parody. How do those bits just continue to exist there on the surface of a black hole? I mean, you have to be so delusional, it's almost amazing. And don't forget to shield those qubits -- I mean if you're going to keep qubits there, you better have them shielded in that black hole very nicely. There's an example of a shield (see above). The qubits are even worse than bits, and we're going to look at that in a second.

Here's the qubits. There are tons of realizations of qubits now. I mean you can find video after video where they're coming up with different versions of qubits. They don't know what to do with the qubits. That's another question and we'll see why in a second. But here's one version that existed for a long time. Actually, the qubit is realized as a classic LC circuit – where, looking at the picture/graphic (see below), L is an inductor (or a coil) and C is a capacitor.

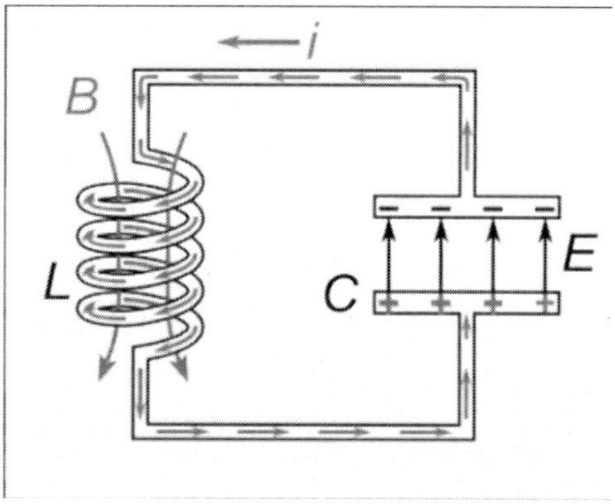

And so, what you're doing is . . . if you notice, you've got the capacitor and it is discharging in going upwards there with our current flow, then through an inductor (coil), and then back again (see below).

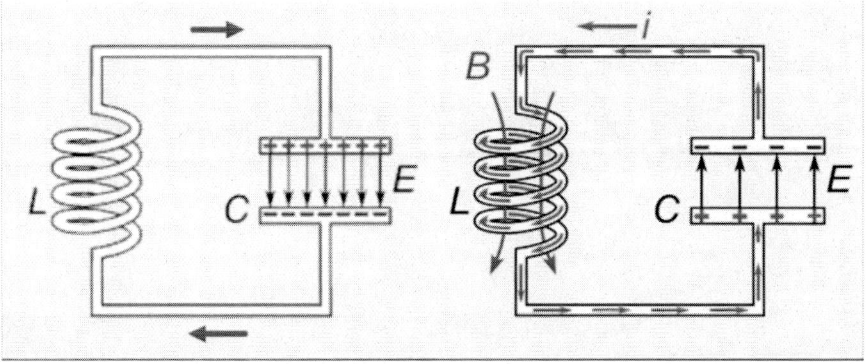

So, this thing will begin to resonate, but gradually dissipate due to the resistance in the circuit, et cetera to zero. Until then it basically discharges back and forth, back and forth (see below).

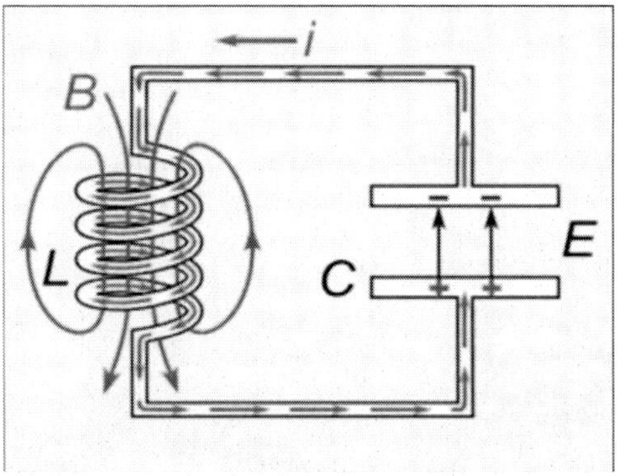

Then what you do is you add a Josephson junction. This is effectively a small resonant antenna.

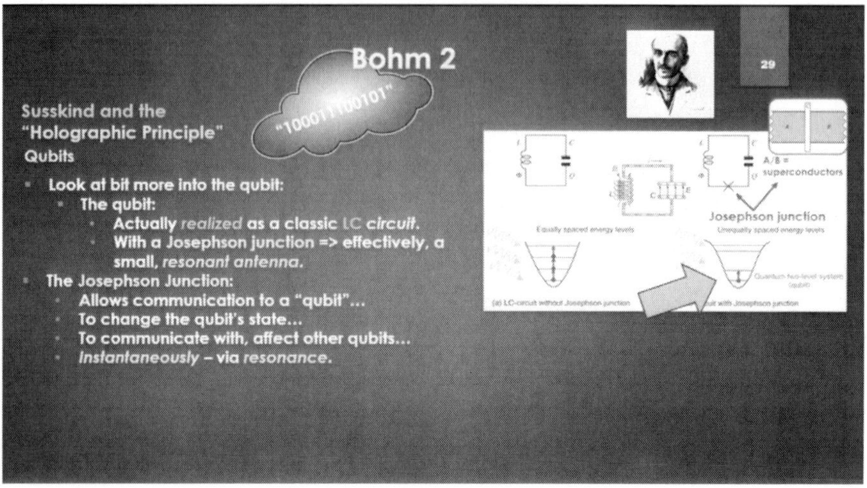

I've got the picture there to the right (see above), you see the Joseph junction. The little X there is inserted into that circuit and what that does is it lets you modulate the state of that LC circuit. So, it allows communication basically to your qubit. Your qubit is the LC circuit. You can change the qubit state to communicate with or affect other qubits instantaneously via resonance with that Josephson junction. Thus, you can basically modulate what level that LC circuit sits in.

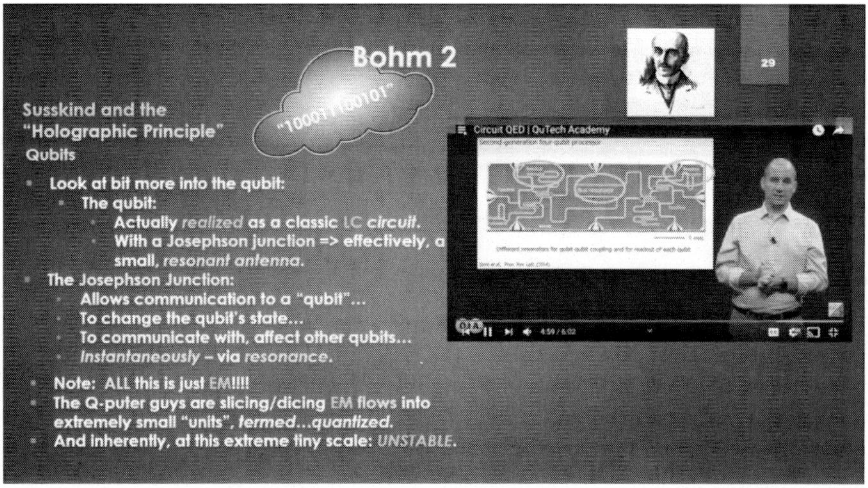

And notice the picture there (see above). This is from one of the courses that my other retired buddy and I looked at on quantum computing. Notice all the little bus resonators, the little resonators, in the diagram there for the qubit-qubit coupling. So, note: **ALL** this is just really **electromagnetism**! The quantum computer guys are slicing and dicing **electromagnetic flows** into extremely small "units," -- which they then **termed quantized**. And inherently, at this extreme tiny scale, they're very **UNSTABLE**.

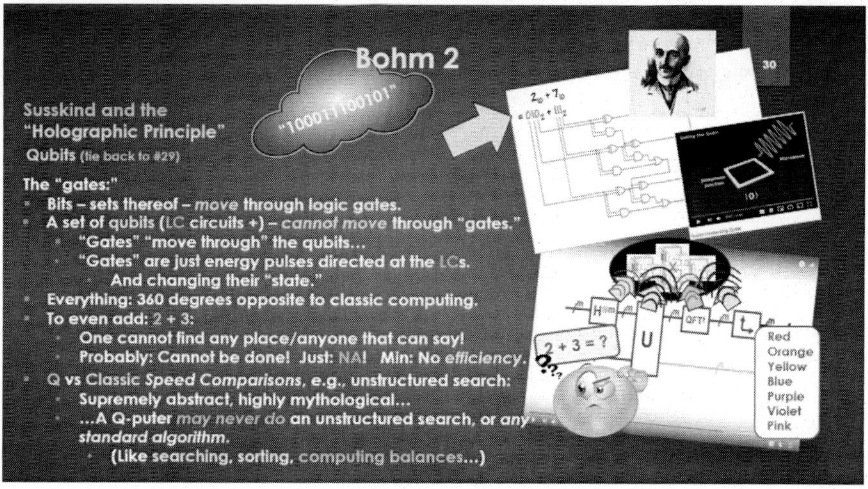

So, the gates in the quantum computer (remember, I am still talking about Susskind's qubits) . . . in a regular computer, the bits move through the logic gates. In that little picture there, I've got a 2 plus a 7, which is a 0 1 0 and a 1 1 1. You move through the logic gates, you move those bits through the various and/or gates, and you'll come out adding them together -- there'll be a 1 0 0 1 at the bottom there, which will be a 9. Okay?

So, the bits move through the gates, but in a set of qubits with LC circuits, you can't move those things through the gates. So, the gates move through the qubits -- gates are just energy pulses directed at the LC circuits and changing their state as such. Okay? So, you're directing these Josephson junction waves at your set of LC circuits -- as my little picture there shows. So, everything is 360-degree opposite to classic computing. So, to even add 2 plus 3, well we can't find any place where anyone can say how they can do that with a quantum computer. Again, I'm sort of going on this qubit thing because you'll find these physicists saying, well really, it's qubits and the whole universe is a quantum computer.

Well, what I'm trying to point out here is not only is the idea that you're going to have qubits sitting on a black hole ridiculous, but the fact that you've got a quantum computer at all is ridiculous. No one can even say how you would add 2 plus 3. We found one guy saying how he could add 1 plus 2, and because we're both old assembler programmers, we realized that all he is doing is . . . it's big kluge of right shifting and left shifting of what would be called assembler bits to achieve adding 1 plus 2 in a so-called quantum computer.

You can get an idea . . . it's very, very difficult to represent a number in this system, let alone add them, let alone sort the numbers, let alone compute balances for a general ledger. It's just not going to happen. It probably just can't be done -- or no efficiency. That's why all you see in the quantum computing world is more new versions of qubits. Fine. Because you don't have to do anything with the qubits. You can make it look like there's progress being made -- oh, a new qubit – but what to do with the qubits? Minor problem. They can't even add 2 plus 3 with their qubits. So, all these quantum versus classic speed comparisons with an unstructured search -- you can find people like Scott Aaronson saying, oh, theoretically a quantum computer could do it faster than a classic computer. It's all baloney – (like searching, sorting, computing balances et cetera).

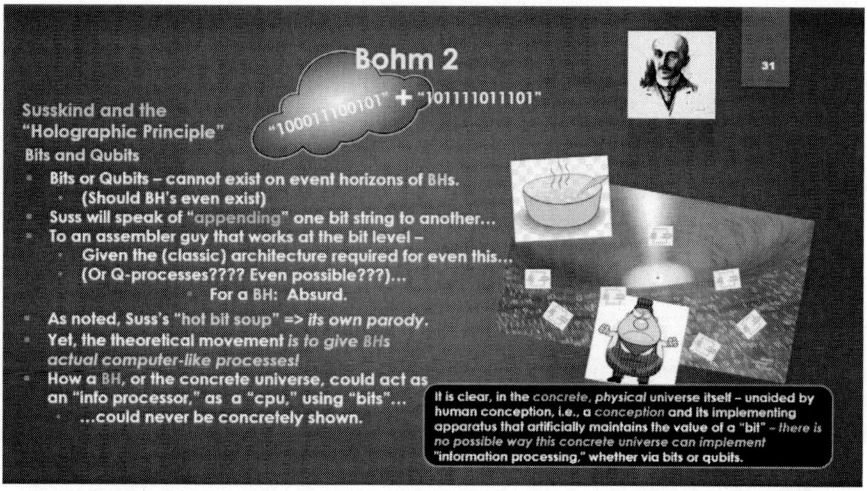

So, to get back to the main point . . . Bits or qubits cannot exist on event horizons of black holes. Susskind will speak of "appending" one bit string to another . . . Given the classic architecture required for even this . . . (it's an incredible architecture) – one can get an idea from that picture I showed -- just adding 7 plus 2 to get 9 -- the structure that's needed -- rigid structure that doesn't move -- isn't wobbly and disintegrating -- this is insane -- that you're going to add, you're going to pin one bit to another. I mean it is absurd. So again, as I noted, his "hot bit soup" is its own parody.

Yet, the theoretical movement **is to give black holes actual computer-like processes!** How a black hole, or the concrete universe, could act as an "information processor," as a "CPU," using "bits" or qubits. . . you could never concretely show that.

It is clear, in the concrete, physical universe itself -- unaided by human conception, i.e., a conception and it's implementing apparatus that artificially maintains the value of a "bit" – there is no possible way this concrete universe can implement "information processing" whether via bits or qubits. This is all a lala land.

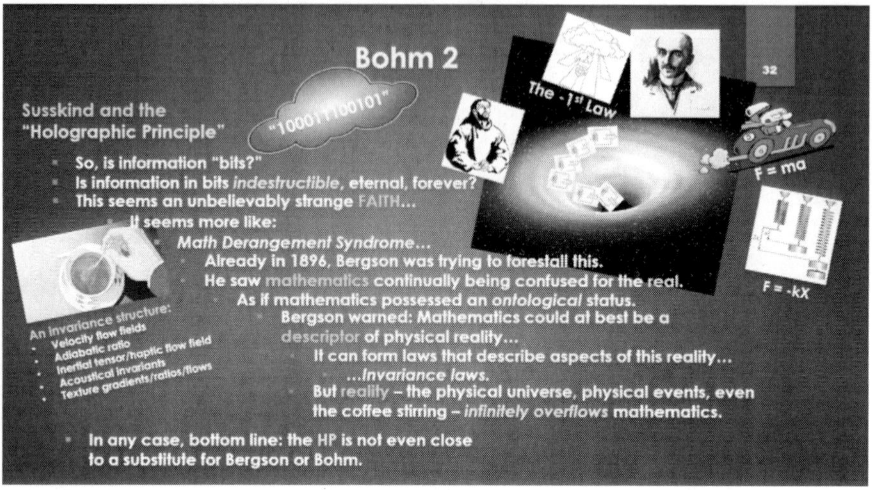

So, is information "bits"? Is information in bits indestructible, eternal forever? Well, this is just a faith. This is an unbelievably strange FAITH. It seems more like Math Derangement Syndrome. Already in 1896, Bergson was trying to forestall this. He saw mathematics continually being confused for the real. As if mathematics possessed an ontological status.

Bergson warned: Mathematics could at best be a **descriptor** of physical reality. It can form laws that describe aspects of this reality -- for example, the invariance laws I showed on the coffee stirring, or f = ma, et cetera. But he said the physical events, even the coffee stirring, infinitely overflows the mathematics --which is true. In any case, **bottom line: the Holographic Principle is not even close to a substitute for Bergson or Bohm**. And if you think it is, the burden of proof is on you.

Pryzdia:
Yes. Are you wrapping this up?

Steve:
I think this is probably the last of the Susskind and Holographic Principle, yeah.

Pryzdia:
Okay. Let me, let's . . .

Steve:
Let's peek forward one more time.

Pryzdia: Go ahead.

Steve:
Yeah, we're going to go to reconciling Bergson with Bohm.

Pryzdia:
Well, let's just summarize what we just saw because the reason I asked you to make sure we cover this with regard to a Bohm-Bergson discussion, is because if you go to the internet or if you go to the published literature, it's just everywhere – the Holographic Principle of Susskind is everywhere. And what always has been bothering me, and this has been for years now, is . . . as you know, I'm more involved in the consciousness studies field -- I'm not a physicist, I'm not a mathematician -- so a lot of my research and reading takes place within the field of consciousness studies or on the nature of consciousness and whatnot.

And "more than a few" people writing in that realm tend to blur Bohm's notion of information with the whole idea of information as "bits." That's just **not** Bohm. And so many people are blurring Bohm and Susskind together like that, and it just drives me crazy. As you said: "Where is meaning?" This is all symbolic language. What does this have to do with nature?

Steve:
Well, in my metaphor, it's, it's slamming tofu blocks on top of tofu blocks. None of the tofu blocks have been actually examined. We just take General Relatiity. We just take black holes and information on black holes. We just take Special

Relativity. We just take a bunch of quantum mechanics. Throw it all together -- and we have a cosmic view of the universe. But none of those tofu blocks has been adequately examined. I have said ad nauseum now: if you more closely examine the tofu block called Special Relativity, it dissolves into a pile of whatever tofu blocks collapse to [laughter], if you heat them up and melt them – a pile of goo, I guess.

Pryzdia:
I just wanted to say that I'm glad we brought this out because, as I've said before, people talk about information going back and forth from an explicate order to an implicate order, but then they refer to that information as "bits" and then they refer to the black holes, and it's just frustrating. So . . . thanks for sharing.

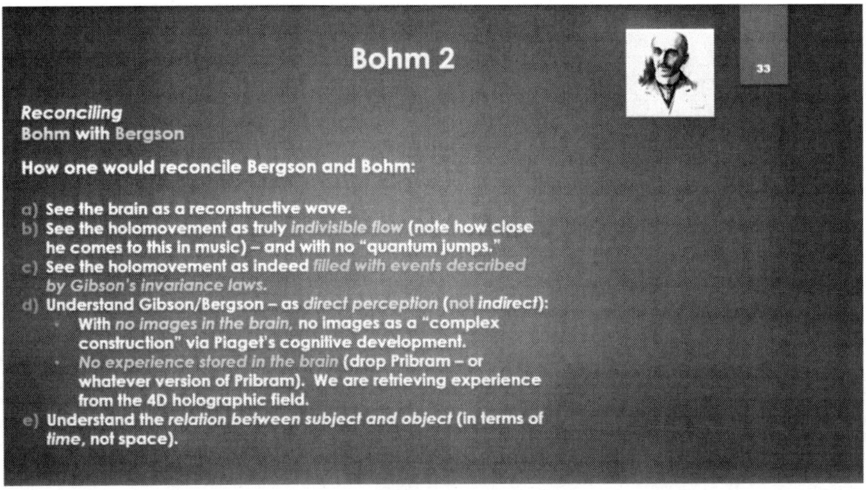

Steve:
OK. We have one more phase here – we are going to try to do a little bit of reconciling with Bohm. Now the problem here is that most of my reconciliation is focused on where Bohm would have to move to hit with Bergson. How one would reconcile Bergson and Bohm:

A) He would have to see the brain as a reconstructive wave -- passing through his holographic field.

B) He would have to see the holomovement as truly **indivisible flow** and Bergson's concept of duration. He comes close to this in music, but again, he'd have to do away with the "quantum jumps" which is another . . . I don't know if you and I had a chance to talk about this -- it would take us off to another topic we could have put in – namely Schrodinger's 1952 paper where he just destroys the notion of quantum jumps. He doesn't leave anything left of it. And people don't understand this. You know?

Pryzdia:
Yes, Bohm adhered to the quantum jumps notion.

Steve:
Right. I don't know if we talked about this, but as something moves from his generative implicate order from state to state, he's trying to allow for that jump from state to state being a quantum jump. And he goes through all that . . . I think we mentioned this . . . how he tries to say that I still have indivisible motion, but there can be quantum jumps in the indivisible motion -- and that becomes such a Zen koan that no one can deal with it. At least I think so. I think he needs to look at Schrodinger and realize that the quantum jump thing may be completely illusory.

C) I think he needs to see the holomovement as indeed **filled with events described by Gibson's invariance laws**. Again, remembering that even Gibson's invariance laws, as we noted, don't have a positive reality – you're not generating these events via mathematical laws. The events are simply being described by the mathematical laws, to some degree.

D) I think he needs to understand Gibson's/Bergson's notion of **direct perception** -- because he's basically, as we noted, an **indirect realist**. He still has the vision of the world in the head or in some mental space -- some perceptual space as Joscha Bach would say -- that is, some completely unanalyzed, unspecifiable region that he could never truly come to grips with concretely. In other words, he's got to get the images in the brain out -- no images as a complex construction via Piaget's cognitive development, as we noted. Again, he has to come to the realization that there is **no experience stored in the brain**. He has to drop Pribram or whatever version of Pribram . . . you and I discussed this -- because I think Pribram has multiple versions, and I don't know which is the real Karl at some point. But I don't think he (Pribram) ever leaves the notion that experience is stored in the brain -- in some incredible holographic redistribution, redistributed memory. I don't think he ever leaves that.

But rather we are retrieving experience from the 4D holographic field -- retrieving experience from the holomovement. It would've been just so interesting if Bohm would've known this version of Bergson and tried to reconcile.

Pryzdia:
Oh yes. I agree, absolutely. Did I tell you . . . I think I mentioned this to you. I was in a conference last year with Basil Hiley. Who else was in there? Some of the people who had worked with Bohm.

Steve:
David Peat?

Pryzdia:
No, Peat had passed away already at that point. But anyway . . . Oh I know – it was Paavo Pylkkänen. He was in there too.

Steve:
OK. Oh yeah.

Pryzdia:
I asked both of them . . . I said, I'm working with this guy, Stephen Robbins and he has this theory that Bergson had a holographic theory . . . and I said, "Hey, did Bohm read Bergson?" And Basil Hiley got all excited and said . . . well, first he said, "Don't ever ask me who Bohm read because whenever you asked Bohm that question, he would answer that he had not read so-and-so, but then you find out later that he had." Now, a lot of Bohm's thinking was very original, but with regard to Bergson, Hiley said, "Well, he must have read Bergson, because I did, and I'm sure I talked to him about Bergson." And I said, well, did he read Whitehead? He goes, "Oh, not really." A lot of people think that he did, and there are so many similarities . . . so it's kind of reverse of what the popular literature is saying. But anyway, my point is that it would've been fascinating if he would have meditated more on Bergson because he does not come out and discuss his work at all.

Steve:
And the insight that Bergson was indeed holographic probably hadn't hit them either. I mean they had . . .

Pryzdia:
No, no, no, no. You're right.

Steve:
There's only one guy pushing that around, and that's me.

Pryzdia:
Yeah, you're right – nobody had gotten that. But another point though, Steve is . . . I think I pointed this out to you, and then I think you did a little research yourself after I

mentioned it to you. . . I pointed out that Bohm was fascinated by Gibson. And in that book that he wrote on Special Relativity, which came out in, I think the mid-1960s, 1965 or something like that – he has a whole Appendix on "Physics and Perception" where he goes into Gibson. Now, I've always been curious . . . what did he pull out of Gibson? He obviously saw a lot of value in what Gibson was doing, but he doesn't discuss him much with regard to the particulars. You know? So, that's another one of those unfortunate things. I think both . . .

Steve:
And again, as I have also been trying to drive home to people -- Gibson fits within Bergson's framework – he only makes sense in the holographic framework. You have that brain resonating away to this structure of information in the external field. Somehow there's an image from that resonance and that resonance is actually going to have to be taken as a reconstructive wave. Again, had he just had that key, it would have been so interesting.

Pryzdia:
Yes.

Steve:
So anyhow . . . I have a couple more here -- one more little slide and one point left to do.

E) Understand the **relationship between subject and object**. Bohm has no concept that this is even a problem – that subject versus object is even a problem -- that you have to have something in your framework, in your philosophy, to deal with that.

Pryzdia:
Can I comment on that? I see where you're coming from. But

I did want to say to you at some point -- I'm glad it's coming up now – Bohm did a lot on the topic subject and object, particularly with regard to his conversations or his dialogues with Krishnamurti. There was a lot of meditating done on the topic. However, like I said yesterday in your presentation, I do see a tendency in Bohm to put a little more value on the order of space as opposed to the order of time. And I think that's your point here, correct? I mean . . .

Steve:
Right, right, right. Yeah. And I'm thinking, well, yeah, my corpus of Bohm's work is *Wholeness and the Implicate Order* and some of the material that Hiley and Peat wrote. But yeah . . . Krishnamurti . . . that whole set of discourses is your area.

Pryzdia:
You have to check those out, you really do.

Steve:
I could understand where that would be a natural context for them to discuss those ruminations, shall we say. Yeah. Well, moving on, I have one more point (see below).

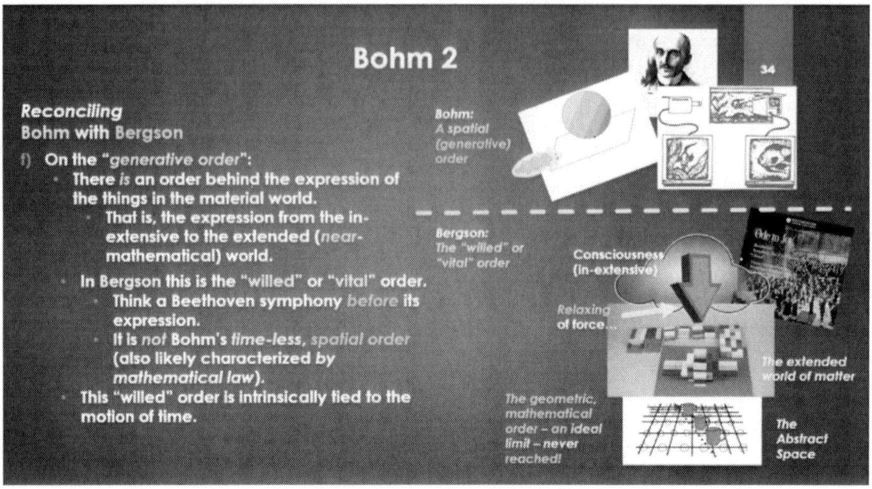

F) The "generative order." So, we know that Bohm has, as we discussed, what I would consider **a spatial generative order** -- symbolized by that sphere, which is sort of a static order, creating another order in that 2D sphere where it looks like there is time.[7] Of course, something's has to move that sphere, and that's where we mentioned that Bohm becomes a little bit incoherent.

With regard to the fish tank example, inside the fish tank there is a dynamic change in that order that is being projected on a 2D plane, namely two different views of the fish. I mean, you could freeze the fish tank and then you could do the equivalent of the ball -- you could have someone move the fish tank back and forth [laughter] or rotate it around and the fish would change. But then again, you're back to the infinite regress. Somebody's moving the fish tank or somebody's sliding the fish tank back and or rotating it around to get that change. He always ends up back at something that's generating time and something that's a continuous order. But

---

[7] As was shared in the previous talk ("On Bergson and Bohm, Part 1").

anyhow, there **is** an order behind the expression of things in the material world in Bohm but also, as we'll see, in Bergson. That is, for Bergson, there is the expression from the in-extensive to the extended (or the **near**-mathematical world). So, in Bergson, this order is what he terms in *Creative Evolution*, the "**willed**" or "**vital**" order. And so, you have consciousness (the in-extensive) **relaxing**, so to speak, to the extended world of matter. The geometrical or mathematical order is an **ideal limit – never** reached! Okay. And then the cloud there, the in-extensive, that order . . . the best analogy that Bergson would use is that it is like a Beethoven symphony before its expression in the physical world as a nice set of written out notes on pieces of paper. But it is not Bohm's timeless spatial order which I think he thinks is probably characterized by mathematical law, the laws of physics -- because as we noted, they have no positive reality. Again, they're just an ideal limit never truly reached. And finally, this willed order of Bergson's is intrinsically tied to the motion of time -- time is a force. So that, to sum up, is kind of the difference between the two notions of order. And they're both there. They both have it, but one is intrinsically involving time and duration and more of a dynamic order than the other. The Bohm spatial, more static order is nevertheless going to require something that moves it into expression.

Pryzdia:
He falls back on a generative order. And don't forget about the quantum potential . . . Is there anything else you wanted to say about this slide?

Steve:
Nope, I think I've rung my little mental dish rag out here on this whole thing. [laughter]

Pryzdia:
Well, I think we did a pretty good job of fleshing out this . . . again, we call it our Bergson-Bohm comparison. Because, as I've said, I've just spent a heck of a lot of time meditating on trying to **integrate** the two. You know? How do you integrate Bergson and Bohm? Because there's such brilliance in Bergson and so much of it . . . it would have been, as you just said, it would have been fascinating if Bohm would have been meditating on what Bergson was saying. So, your last few slides did a great job -- these are the areas where "only if" -- because there's such potential to integrate the two. And I think it all does boil down to that whole notion that **time is creative itself** -- that's what seems so important, for me anyway -- that's what I'm finding myself meditating on. **How does time create novelty?**

Steve:
Right. This is a major profound subject. As I've noted several times, I haven't really . . . I didn't feel like it was really my mission in this life to dwell on . . .

Pryzdia:
I'll make it mine. [laughter]

Steve:
Or deal with very much.

Pryzdia:
All right Steve This was great. Anything else you wanted to say with regard to concluding things, summarizing, wrapping things up?

Steve:
Nope. I think we can really sum it up with: "if only" -- if only Bohm had understood these aspects of Bergson. It would've been amazing. That would've been interesting

Pryzdia:

Here's an observation, and I guess it's a question as well because again, as you know, I've been meditating a lot on this notion of pulsating matter and pulsating time, pulsating duration, time is discontinuous as opposed to continuous. And I was wondering, Bohm himself, like you, felt that science, for instance quantum physics, needed to get to the bottom of what's actually taking place in the natural world as opposed to just relying on mathematical equations as somehow generating cosmos. He was very much against that. Apparently when he would lecture, he would lecture at one place just using nothing but narration, talking about meaning. And then he would go the next day or the same day to another university [focusing on the same subject] and use nothing but mathematics. And, of course, when he would use nothing but mathematics on his board as he's lecturing, all the students just ate it up. But the other lecture where he did not use as many mathematical equations, people would say things like "oh, this is mushy, this is sloppy." He dealt with that his whole life apparently. But my observation is this: it seems that quantum physics is, to use Bergson's language, a creation of the "intellect" – to be distinguished from "intuition." With intellect, everything is chopped up and you wonder what's real here. Have you ever thought about that? Well, I know you have – you have done a pretty good job of pointing out some of the limitations of quantum physics – for example, the nature of a photon -- we didn't cover that either.

But anyway, I've been thinking a lot about Bergson's distinction between intellect and intuition. Again, using Bohm as an example, Bohm obviously would have resonated with Bergson's notion of intuition. I mean, for him, reality is nothing but movement. But at the same time, in my opinion, he had a tendency to fall back more on using intellect. And I think I've told you that when he would dialogue with Krishnamurti, sometimes Krishnamurti would quiz him -- there was something going on with time. As you just pointed out in this presentation, Bohm had those notions of "moments," those "instants" of time. And often Krishnamurti would kind of catch him on that. In my opinion, Bohm was always . . . he always wanted to **think and then see something.** And Krishnamurti would attempt to communicate something to the effect of – **what if you were to see first and then think**? And by think, I mean intellectually think -- intellect. That used to frustrate Krishnamurti a bit – I gathered – because Bohm would always fall back on what's going to happen the next moment. And Krishnamurti would say, why don't you **SEE** and find out? [laughter] Now I'm paraphrasing here of course -- Krishnamurti would never speak to him like that, but do you get my notion there? He really did view time, as you're pointing out . . .he used moments as a way to understand it – instants in time. And I think that became an Achilles heel for Bohm, a little bit.

Steve:
Right. If we took that to the Bergson case, Bergson would've been the epitome of what Krishnamurti was counselling Bohm to do. That is, when Bergson started, his whole thing was "wow, when you really look at your experience, the whole notion of abstract time and space dissolves into an obviously complete artificiality." It's this description of experience that he has, that describes duration, that makes you wonder if he was on LSD or something.

Pryzdia:
Yes. [laughter]

Steve:
But yeah -- it was the direct observation of experience, which is what he counseled. I mean he would've probably been sitting right next to Krishnamurti counseling Bohm. The two would've been speaking roughly the same language. An interesting thing to think about . . . did you tell me?. . . Yeah, you told me that Krishnamurti did sit in some lectures of Bergson.

Pryzdia:
Apparently when he was a young man in France, there is . . . I believe this appeared in a biography that came out fairly recently. . . but yes, apparently there's evidence that he sat and listened to Bergson in France when he was a very young man. So yeah, if that's true that's crazy -- because I see all kinds of parallels between what Bergson was saying and what Krishnamurti was saying.

Steve:
So, we almost have Bergson in the form of Krishnamurti talking to Bohm.

Pryzdia:
Yeah, that's one way to look at it – that's a good point. So, all right. Well, thanks Steve.

Steve:
Hey, it's been fun. [laughter]

Pryzdia:
You and I are weird guys – it has been fun for me as well. So, thank you for your time.

Steve:
Yeah, it's been interesting. It's been generative -- it's been a generative order here. [laughter]

Pryzdia:
Well, I'll talk to you soon, okay?

Steve:
Okay.

Pryzdia:
I'm going to stop recording. We will stop this off here. Just stay in touch.

Steve:
Will do.

Pryzdia:
All right. See you soon.

Steve:
See you later. Okay, bye-bye.

Afterword

It has almost been three years since Steve and I did our series of dialogues. I noted in the Introduction that it was my intention to share some additional thoughts with regard to what Steve shared; and thus, I would like to take this opportunity to comment a bit more here. I still believe that the chapters of this book are rich in detail and thought-provoking content. And... I am still asking this question: "Why is the work of Stephen Earle Robbins (books, multiple articles, and, as of the date of this writing, now over 100 YouTube video presentations) still being virtually ignored by the academic community?"

Yes, I am aware of Steve's tone and the fact that he does not fully adhere to scholarly decorum – this is regularly pointed out to me by the scholarly friends/colleagues with whom I share Steve's work. But... this was a Captain in the U.S. Marines who fought in the jungles of Vietnam. He learned to hone certain skills; and while in combat, he continually needed "to be practical" – he continually needed to adapt to his environment and focus on only that which was absolutely necessary, and valuable. I have always assumed that Steve's ability to cut to the chase, distill the gold (that which is valuable) and necessarily and effectively discard that which is not necessarily valuable was carried from his days in combat into his career as a scholar/writer. Thus... I am personally willing to accept the "grenades" he throws at orthodox views (held within for example: physics, consciousness studies, ecological psychology, etc.). Indeed, in my opinion, we need more thinkers like Stephen Earle

Robbins — human beings who take the time to actually "look under the hood" (as he likes to say) — and bring to our attention more than a few emperors without clothes. I agree with Steve: the extent to which scientists and philosophers (and "thinkers" in general) are so thoroughly hypnotized (and thus incapable of suspending their assumptions — thereby rendering them utterly incapable of actually listening to points of views that differ from the ideological perspectives that they have so thoroughly identified themselves with -- and, for many, staking their professional careers on) — is truly shocking. Not an edifying spectacle.

How does one **integrate** the holographic vision of Bergson with the holographic vision of Bohm? Meditating on this particular question has been a fascination of mine for quite some time now. Indeed, it was this very question that motivated me to suggest a series of dialogues with Steve in the first place. Can such an integration even be done? I have thought long and hard about this. Some days I would think it can be done, and I outlined ways in which this is possible. And on other days I would just give up and decide that it cannot be done – that both visions are very strong in their own unique way and need to be honored as such. As I have said, there are multiple similarities between both visions – and multiple differences. And each vision has its own set of strengths and weaknesses. In the end, I do think that such an integration can be done; but there are many subtle details that will have to be shared and explored in order to put forth a coherent picture of how this is possible; and this is not the time or space to do so. Thus, below are some very general "observations." They are a brief collection of not so much "conclusions" but rather "proposals" that come out of my dialogues with Steve.

Just like the work of J.J. Gibson needs to be placed within the context of the holographic vision of Henri Bergson (I agree with Robbins here), so too does Bergson's work need to be placed within the context of the work of David Bohm (I suspect Robbins would disagree with me on this). In my opinion, after meditating upon this topic for quite a while now, I believe that Bergson and Bohm can be integrated as long as one places Bergson within Bohm. More to the point, it could be said that Bergson is **already** working within Bohm. Certainly, Bergson does not **need** Bohm. But since Bergson's philosophy does include a holographic vision (and again, I am quite thankful for Robbins for discovering or pointing out the existence of this vision), I do feel that Bergson's philosophy can become more coherent (or more "complete") once placed within Bohm's physics (and philosophy). Again, a description of a Bergson-Bohm integration will have to wait for another time. Right now, I am settling with a simple Bergson-Bohm comparison, or a Bergson-Bohm comparison/contrast. A brief list outlining a few points is in order.

Thus, for clarity sake, when I say that Bergson is already working within Bohm, what I mean is that Bergson's holographic vision does not fully work (or is at least less coherent or complete) without positing the existence of a Bohmian "generative order." As I alluded to when chatting with Steve, **this generative order is already implied in Bergson's work** (i.e., duration "unfolds" extensity [there is a movement from the "intensive" to the "extensive"]; pure memory "unfolds" memory images; dynamic schemes "unfold" actions, etc.). For all the talk in Bergson of tendencies and the relaxing of tendencies, we still cannot help but notice the language used – the "unfoldings." I see a difference between an "unfolding" and a "relaxing of a tension." And the notion of "tendencies" that Bergson was so fond of (owing in no small measure to his attraction to the Stoic tradition) is, at least in my mind, seemingly necessarily tied to memory. And one of the points I made in my dialogues with Steve is that if duration/tendency is so closely tied to memory, how then can we ever have true creation/novelty (and not simply invention) -- without a generative order? In my mind anyway, **invention depends on memory** (the storage of information) while **creation does not depend on memory**). For this (creation) we need "the timeless" order (which is also a requirement for "enlightenment" as typically defined in the Buddhist and Advaita Vedanta traditions [8]). In my opinion, Bergson's work needed to make room for the timeless order. BUT, as David Bohm repeatedly suggested toward the end of his career/life, we do not want to suggest that the timeless order is some kind of "absolute ground." As Bohm suggested, we want to move away from meditating on the order of time v/s the timeless order – we want to move away from meditating on the continuity of time v/s the discontinuity of time (see for example, *Science, Order, and Creativity* (1987, 2000) and *Physics and the Ultimate Significance of* Time (1986) – and the works of Basil Hiley). [9]

---

[8] In these traditions, enlightenment was not simply an illusion/delusion (the product of a limited "intellect"/thought).

[9] It is worth noting that Ian McGilchrist is a contemporary writer who sounds very much like David Bohm did toward the end of his career; see, for example, *The Matter with Things: Our Brains, Our Delusions, and the Unmaking of the World* (2 Volumes),

For David Bohm, it was most important to pay close **attention** to the **interplay** between the order of time and the timeless order – and to **understand** this **relationship**. Working solely within Bergson's holographic vision, we do not really have this opportunity -- as there is, at the very least, a de-emphasis of the ontological value of the timeless. And with regard to the temporal metaphysic v/s the classic metaphysic: I would simply like to share that if Bohm is working within the classic metaphysic, he is certainly doing so with a conscious awareness of its limitations! [10] Further, taking Robbins' critique of the classic metaphysic to heart, might we be able to conclude that despite its limitations, Bohm makes the metaphysic work? That is, as Robbins often shares, the similarities between Bohm's general metaphysical vision to that of Bergson is not to be lost sight of – "Bohm is so close to Bergson." In my opinion, the key link to both holographic visions is the "generative order."

It was in Dialogue #3 where I first asked Steve the question: "... why can't we look at Bergson's duration as a generative order? [...] What do you think of that notion?"

2021. McGilchrist refers to both Bohm and Robbins in his book – and makes it very clear that he is a HUGE fan of Bergson: "Yes, if I was asked, which philosopher in the last 100 years it is most important to read, it would be Bergson."

[10] This amounts to, by the way, of an awareness of the **limitations of thought** itself – a MAJOR area of exploration of Bohm and Jiddu Krishnamurti; see, for example, *The Limits of Thought* (1999) and *Thought as a System* (1992). Bergson too was very much interested in the limitations of he called the "intellect."

There was then a long pause before Steve said: "You have to be careful. […] Because you can't imply that 'it's all there' in that order and it's just unfolding. […] That's totally non-Bergsonian. Bergson would say, 'that's just teleology.' You're just unfolding toward an already defined goal. And that goal isn't defined in reality." I then went on to say that it is a "subtle issue" and we shifted to a new topic. I will say now though that while Bohm was attracted to teleology, it was never as extreme as is suggested by Steve. That is to say, the generative order is truly **generative** for Bohm.

The new topic we shifted to was "order" itself; and here Steve seems to insinuate that Bohm suffered from what Steve calls "mathematics derangement syndrome." This is not the case at all. Bohm was hyper-focused on using mathematics to **describe** the nature of reality; his mathematical explanations were not grounded in the belief that mathematics is an ontological ground – I do not think he would be impressed with the conclusions of writers today who seem to put forward the notion that mathematics generates the cosmos/nature. As I alluded to in Dialogue #3 and Dialogue #4 though, the notion of order was very fundamental for Bohm – and, at the same time, it is almost something that cannot be defined. I mentioned in Dialogue #4 that the fundamental nature of order for Bohm is very much tied to the work he did with Jiddu Krishnamurti; and how sometimes Bohm, if pressed to define order (and sounding like a mystic) would occasionally say things like "order is perception."

In my opinion, the book that David Bohm wrote with David Peat – *Science, Order, and Creativity: A Dramatic*

New Look at the Creative Roots of Science and Life (1987) is a masterpiece; it is here that order is explored in some detail. And it is here where we read of the importance of understanding that order is context-dependent. Bohm, addressing the question of whether or not order is simply in the mind, states: "To attempt to attribute order solely to the object or to the subject is too limited. It is both and neither, and yet something beyond all this: a ***dynamic process*** [italics mine] that involves subject, object, and the cycle of perception-communication that unites and relates them (115)." Interestingly, in the next paragraph, Bohm comments on the inadequacy of the Cartesian grid: "... the Cartesian grid or coordinate system, has dominated the basic order of physical reality for the past three hundred years. In many cases the Cartesian grid worked well, in the sense that it led to a coherent activity and thus corresponds to reality fairly well. However, in this book, the general appropriateness of the Cartesian description is called into question (115)."

As I pointed out in my dialogues with Steve, **perhaps we should consider not separating "order" from "movement**." This notion dawned on me while in dialogue with Steve, but I struggled with what such a notion could really mean. Obviously, we can go with the notion that order is first (or more primary), then movement. Or we can go with movement (i.e., as time – which is Steve's preferred position) is first, then order. OR we can go with order **IS** movement – and this would point us in the direction of Bohm's "holomovement." I believe that one of the most interesting sections of my dialogues with Steve is located in Dialogue #3 where the holomovement is addressed, and where Steve shares his negative assessment of the term or at least its function within Bohm's physics/philosophy. Steve's main point is that Bohm's generative order is too static (he refers to it as a "static spatial order"); and he illustrates his point by using the image of a three-dimensional sphere moving through a two-dimensional plane. [11] The holomovement is a fascinating (and subtle) notion – as is its distinction from the generative order. The notion has been explored by various writers over the years; and it needs to pointed out that toward the end of his life Bohm did address possible limitations inherent in using the term – he preferred substituting it with the term **"holoflux";** see, for example Mark Schroll's "Understanding Bohm's Holoflux: Clearing Up a Conceptual Misunderstanding of the Holographic Paradigm and Clarifying its Significance to Transpersonal Studies of Consciousness" (2013).

Schroll writes: "*Flux* refers to a change in state rather than movement in time or place (153)." And his paper on Bohm's use of the term holoflux is quite interesting; it wound up being very useful to me with regard to my attempt to integrate the holographic visions of Bohm and Bergson -- as well as allowing me to flesh out Steve's objection to Bohm's use of the holomovement and the generative order. In essence, I felt that it allowed me to get past Steve's use of the spatial metaphor of a moving sphere. The moving sphere was

---

[11] As I pointed out in a footnote, to my knowledge (and I certainly could be wrong here) David Bohm never used this spatial metaphor of a moving sphere to help understand the dynamics of the holomovement/generative order. But it is interesting to note that Shell Joye is another writer prone to using spheres in conjunction with Bohm's work in her writings on the nature of consciousness.

not the effect of a cause that was static (Steve's image of the generative order) but rather Bohm's generative order is a "force" – in the sense that Bergson used the term: for Bergson, a force is always ultimately motion itself. As Steve writes in *The Challenge to AI: Consciousness and Ecological General Intelligence* (2024), "Force is no more absolute than the movements: it cannot serve to distinguish real or absolute motion. It is not a 'cause.'" Quoting Bergson, Steve writes: "It is in vain, then, that we seek to found the reality of motion on a cause which is distinct from it: analysis *always brings us back to motion itself* (215-216)."

If I am reading Schroll correctly, we have, for Bohm, the following: the holomovement takes place within the generative order (it is the process of injection, projection, reprojection) – which is made possible because of the holoflux. Shifting to Bergson: to be sure, Bergson's "holofield" (as Steve likes to call it) is in motion, but before there are motions of objects, there is the flowing holofield itself – there is a holoflux. Again, "*Flux* refers to a change in state rather than movement in time or space." And here is Steve (in the third video of his YouTube series "Bergson and Holographic Theory of Mind") commenting on Bergson's holofield: "We must view the ***whole*** as changing, he [Bergson] argued, as though a kaleidoscope. The 'motions' of separate 'objects' become ***changes or transference of state*** within this global transformation or motion (like waves in the moving sea). It is a transformation with an inherent simultaneity, and it is *indivisible* -- like a melody (Slide 43)." The holoflux applies to both Bohm and Bergson.

As many know, the aim of Bohm's philosophical work was to understand the nature of movement. [12] And while he was not officially counted as an official "process philosopher" within the Western philosophical tradition, he very well could be. Bohm, like Bergson, was focused on "process." Both men were focused on the movement of information. Which thinker has the more eloquent holographic vision? For me it is actually a toss-up. In this Afterword I am only mentioning the notion of possibly conceiving Bergson's time as a Bohmian generative order so that we can bring in the timeless order (and "creation") in a coherent way; for again, I feel that this is necessary in order to complete Bergson's vision. Is the "timeless" simply the effect of "spatializing time"? I think not.

And again, one of the big questions for Bohm was: "What is the **interplay** between the order of time and the timeless?" There is an interaction between our thinking in the order of time and the timeless order. What is the nature of this interaction, and what does it mean? There is a **meaning** here that matters! Again, for Bohm, **"meaning IS being."** And, even more important to grasp, meaning is grounded in a **MYSTERY**; that is to say, there is no absolute

---

[12] This aim has its roots in an experience he had as a boy walking in the woods: "On one of these walks, he encountered a stream that could be crossed only by means of a few stepping stones protruding above the water. Bohm tried to work out in advance how he would place his feet on each stone, but once he got started, he found he had to keep moving and his plans were of no use. This experience precipitated an insight, one he often remarked upon in later years, regarding the nature of flowing movement in all facets of life." (Moody, 6)

ground. As Bohm tirelessly pointed out, one needs to **"suspend one's assumptions"** about "what is" and "what is not" – which opens us up to the possibility of insight. For Bohm, "reality is what we think about" (this is both our "being" and our "becoming" [13]). And for Bohm, true insight involves a SEEING (a kind of attention) that involves the timeless – insight requires the timeless.

Where is our attention? And can we suspend (not suppress) our assumptions in order to have insight – the ability to "see-into" our knowledge (what we think we know)? Which brings us back to the work of Stephen Earle Robbins. Can we take a good look at what he shares? For example, what was the impact of the Special Theory of Relativity on Bohm's work? What Robbins says here is not be lost sight of. Did Bohm "drink the Kool-Aid"? Did the Special Theory of Relativity mess Bohm up? And what do we make of Bohm's indirect realism and his adherence to Pribram? Despite the fact that Bohm was one of the first scientists to pay attention to the work of J.J. Gibson, [14] did Bohm lack a full theory of perception? I greatly appreciate how Stephen Earle Robbins approaches and processes the work of David Bohm -- and Henri Bergson. And I am quite thankful for the dialogues I did with Steve. We both made a few discoveries, which is what dialogues are all about. And so many more questions

---

[13] Which can also include our comprehension of the "becoming of being" and the "being of becoming." See Bohm's "Time, the Implicate Order, and Pre-Space." *Physics and the Ultimate Significance of Time*, edited by David R. Griffin, 1986, pp.177-208.

[14] See, for example, the Appendix of Bohm's *The Special Theory* (1965) – "Physics and Perception."

can be asked regarding the holographic theories of Bergson and Bohm! In closing, I would like to highlight here how important the asking of questions is. I would like to suggest that our scientists, philosophers – all of us – learn how to ask a question. Can we learn how to ask a question without the need to answer it? Let me explain what I mean by sharing the following: Time is a movement. Thought is a movement. Thought is time (or time is thought). But as Bergson pointed out in *Matter and Memory* and *Creative Evolution*, this does not necessarily mean that time is only thought. The trick here is that the movement of time and the movement thought are movements that thought struggles to understand (due to the limits of thought — or Bergson's "intellect"). Certainly, it is the intellect itself that makes Bergson's temporal metaphysics so difficult to assimilate. But still, (and this is an aspect found in the holographic vision of Bohm that is not necessarily found in the holographic vision of Bergson), the question arises: what if human beings begin to place conscious awareness on the **function of thought**? — what if human beings begin to watch what thought itself is doing to them? Could thought then become self-aware? Could thought then become proprioceptive, as the body currently is? Bohm proposes that "the proprioception of thought" could radically transform humanity. For Bohm, in a nutshell, watching what the **"system of thought"** is doing to one can allow "insight" to play a fundamental role in one's life. If we begin to "SEE-into" our knowledge — to SEE what knowledge is DOING, then... well... let us SEE...

I am attempting to highlight here the value and power of attention. Can we attend to the system of thought by suspending our assumptions around Steve's work in order to make some discoveries? I would like to close this Afterword by directing the reader's attention to the Appendix that follows – and its invitation to dialogue with me around the work of Stephen Earle Robbins and what he shared in his dialogues with me. I trust you have enjoyed my interaction with Steve, some of the highlights I share here, and I would relish meeting with you.

Appendix

Why is the Work of Stephen Earle Robbins Important?

I believe Dr. Robbins is being quite humble in referring to "Henri Bergson's Holographic Theory." For this holographic theory of Bergson owes MUCH to Robbins! I am continually amazed by what Robbins discovered in Bergson – and how he assesses recent developments in consciousness studies, AI, perception theory, memory theory, ecological psychology, etc. Below are a few highlights to consider.

What are Robbins' own unique contributions? A short list could include:

· He discovered the holographic theory in Bergson's work – and describes/explains it thoroughly in his papers, books, and YouTube presentations.

· He explains how the perception theorist J.J. Gibson needed Bergson – and how even today the ecological psychology people need Bergson's holographic vision.

· He explains how Bohm does not need Pribram (and his indirect realism/perception).

· He explains how philosophers and scientists and AI theorists need Gibson's direct perception (including what is said about time scales and affordances).

· He explains how consciousness studies people (and again, AI theorists) need to account for the origin of the image of the external world (and thus, in his opinion, solve "the hard problem") by, for instance, adopting the Bergson/Gibson holographic vision -- or perhaps we should call this the "Robbins Holographic Vision"!

· He explains how the consciousness studies people who adhere to a "The Holographic Principle" (e.g., Susskind et al) need a concrete reconstructive wave in order to reconstruct the hologram within their holographic vision.

And...

Although the number of topics that Steve addresses is certainly numerous...

People could possibly benefit by meditating on what Steve says about (my short list is below):

· The importance and power of a "temporal metaphysic" as opposed to what Steve calls the "classic metaphysic" – here he is being helped today by the work of Ian McGilchrist.

· Quantum jumps (and the ignoring of Schrodinger's 1952 paper on this topic).

· How Einstein's Special Theory of Relativity (STR) has been grossly "misplayed" by physicists.

· The sadly halted research that was being done on electromagnetism (EM) (e.g., Tesla, Heaviside, Alexanderson).

· The sad allegiance to Shannon information as opposed to meaning (e.g., for theorists like David for Bohm it is meaning that gives form to matter [meaning "in-forms" matter] – not "bits").

· Steve's challenge to the research and application of Artificial Intelligence (AI) (e.g., see *The Challenge to AI: Consciousness and Ecological General Intelligence* (2024)).

NOTE: (below is Steve's attempt to summarize, in his own opinion, the two key take-aways regarding the significance of Bergson): Steve says in his "Exploring Bergson's Holographic Theory: Key Concepts" YouTube presentation that Bergson gave us:

- The solution to the famous "**hard problem** of **consciousness**."
- The only, concrete alternative to the computer model of mind.

An Invitation:

For those who want to suggest that Dr. Robbins is "wrong," do not hesitate to contact me so that we can gather in order for you to make your case. That is, SHOW how Dr. Robbins is wrong, rather than simply declaring he is wrong. Do not hesitate to contact me if you are interested in taking place in an actual dialogue. I would like to take part in a gathering where people can suspend their assumptions (or at least be open to attempting to do so) so that a discovery may be made – so that something can be learned.

Michael.Pryzdia@asu.edu

Works Cited

Alexander, Ian W. *Bergson*. Bowes & Bowes, 1957.

Allen, Barry. *Living in Time: The Philosophy of Henri Bergson*. Oxford University Press, 2023.

Barnard, William G. *Living Consciousness: The Metaphysical Vision of Henri Bergson*. Sunny Press, 2011.

Bergson, Henri. *Creative Evolution*. 1911. Translated by Arthur Mitchell. Dover, 1998.

---. *The Creative Mind: An Introduction to Metaphysics*. 1946. Translated by Mabelle l. Andison. Dover, 2007.

---. *Laughter: An Essay on the Meaning of the Comic*. 1911. Translated by Cloudesley Brereton and Fred Rothwell. Martino Publishing, 2014.

---. *Matter and Memory*. 1912. Translated by N.M. Paul and W.S. Palmer. Cosimo Classics, 2007.

---. *Mind-Energy: Lectures and Essays*. Translated by H. Wildon Carr. Henry Holt and Company, 1920.

---. *Time and Free Will: An Essay on the Immediate Data of Consciousness*. 1913. Translated by F.L. Pogson. Dover, 2001.

---. *The Two Sources of Morality and Religion*. 1932. Translated by R. Ashley Audra and Cloudesley Brereton. University of Notre Dame Press, 1935.

Bohm, David. *Causality & Chance in Modern Physics*. University of Pennsylvania Press, 1957.

---. *On Dialogue*, edited by Lee Nichol. Routledge, 1996.

---. *The Special Theory of Relativity*. Routledge, 1965.

---. *Thought as a System*. Routledge, 1992.

---. *Wholeness and the Implicate Order*. Routledge, 1980.

Bohm, David and B. J. Hiley. *The Undivided Universe: An Ontological Interpretation of Quantum Theory*. Routledge, 1993.

Bohm, David and David Peat. *Science, Order, and Creativity: A Dramatic New Look at the Creative Roots of Science and Life*. 1987. Routledge, 2011.

Capek, Milic. *Bergson and Modern Physics*, edited by Robert S. Cohen and Marx W. Wartofsky. D. Reidel Publishing Company, 1971.

Combs, Allan. *Consciousness Explained Better*. Paragon House, 2009.

Hiley, B.J. "Non-commutative geometry, the Bohm interpretation and the mind-matter relationship." In: Dubois, D.M. (editor.), *Computing Anticipatory Systems*, AIP Press, 2001, pp. 77–88.

Hiley, B.J. and F. David Peat. *Quantum Implications: Essays in Honor of David. Bohm*. Routledge, 1987.

Gibson, J.J. "A theory of direct visual perception". Edited by Royce, Joseph R. and William Warren Rozeboom. *The Psychology of Knowing*. Gordon and Breach, 1973.

Griffin, David R. *Physics and the Ultimate Significance of Time*. State University of New York Press, 1986.

Gunter, P.A.Y., editor. *Bergson and Modern Physics*. University of Tennessee Press, 1969.

Hirai, Yasushi. *Bergson's Scientific Metaphysics: Matter and Memory Today*. Bloomsbury Academic, 2023.

Jayakar, Pupul. *Krishnamurti: A Biography*. Harper, 1986.

Junior, Olival Freire. *David Bohm: A Life Dedicated to Understanding the Quantum World*. Springer, 2019.

Joye, Shelli. *Developing Supersensible Perception: Knowledge of the Higher Worlds through Entheogens, Prayer, and Nondual Awareness*. Inner Traditions, 2019.

---. *The Electromagnetic Brain: EM Field Theories on the Nature of Consciousness*. Inner Traditions, 2019.

---. *The Metaverse of Consciousness: Mapping the Multiple Dimensions of Reality*. Inner Traditions, 2025.

---. *Tantric Psychophysics: A Structural Map of Altered States and the Dynamics of Consciousness*. Inner Traditions, 2021.

Kolakowski, Leszek. *Bergson*. St. Augustine's Press, 1985.

Krishnamurti, Jiddu. *The Awakening of Intelligence*. Harper, 1973.

---. *Freedom From the Known*. Harper, 1969.

---. *Truth and Actuality*. Harper, 1980.

---. *The Wholeness of Life*. Harper, 1979.

Krishnamurti, Jiddu and David Bohm. *The Ending of Time: Where Philosophy and Physics Meet*. Harper, 2014.

---. *The Limits of Thought*. Routledge, 1999.

McGilchrist, Ian. *The Matter with Things: Our Brains, Our Delusions, and the Unmaking of the World* (2 Volumes). Perspectiva Press, 2021.

Moody, David. *An Uncommon Collaboration: David Bohm and J. Krishnamurti*. Alpha Centauri Press, 2016.

Norsen, Travis. *Foundations of Quantum Mechanics: An Exploration of the Physical Meaning of Quantum Theory*. Springer, 2017.

Peat, F. David. *Infinite Potential: The Life and Times of David Bohm*. Basic Books, 1997.

Pylkkanen, Paavo. *Mind, Matter and the Implicate Order*. Springer, 2007.

Pribram, Karl. *Brain and Perception: Holonomy and Structure in Figural Processing*. Routledge, 1985.

---. *The Form Within: My Point of View*. Prospecta Press, 2013.

---. *Brain and Perception: Holonomy and Structure in Figural Processing*. Routledge, 1985.

Robbins, Stephen Earle. "Bergson and Holographic Theory of Mind." *YouTube*. https://www.youtube.com/channel/UCkj-ob9OuaMhRIDqfvnBxoQ

---. *The Challenge to AI: Consciousness and Ecological General Intelligence*. Mercury Learning and Information, 2024.

---. *Time and Memory: A Primer on the Scientific Mysticism of Consciousness*. Stephen E. Robbins, 2012.

Schroll, Mark. "Understanding Bohm's Holoflux: Clearing Up a Conceptual Misunderstanding of the Holographic Paradigm and Clarifying its Significance to Transpersonal Studies of Consciousness. *International Journal of Transpersonal Studies*, vol. 32, issue 1, 2013, pp. 140-163.

Talbot, Chris, editor. *David Bohm: Causality and Chance, Letters to Three Women*, edited by Chris Talbot, Springer, 2017.

---, editor. *David Bohm's Critique of Modern Physics: Letters to Jeffrey Bub, 1966-1969*. Springer, 2020.

Talbot, Michael. *The Holographic Universe*. Harper, 1991.

Wilber, Ken. *Finding Radical Wholeness: The Integral Path to Unity, Growth, and Delight*. Shambhala, 2024.

---, editor. *The Holographic Paradigm and Other Paradoxes: Exploring the Leading Edge of Science.* Shambhala, 1982.

About the Author

Michael R. Pryzdia is an experienced teacher, writer, speaker, and organizational consultant. He has worked with Fortune 500 corporations, non-profits, and governmental agencies. He is a Professor in the College of Integrative Sciences and Arts at Arizona State University whose expertise lies within the field of consciousness studies.

Made in United States
North Haven, CT
04 November 2026

21771715R00116